DATE DUE

~~JA 4 '94~~			
AP 22 '94			
~~OC 28 '94~~			
AP 28			
~~AP 26 02~~			

DEMCO 38-296

Oklahoma Notes

Basic-Sciences Review for Medical Licensure
Developed at
The University of Oklahoma, College of Medicine

Suitable Reviews for:
United States Medical Licensing Examination
(USMLE), Step 1
Federation Licensing Examination (FLEX)

Oklahoma Notes

Biochemistry

Second Edition

Edited by
Thomas Briggs
Albert M. Chandler

With Contributions by
Thomas Briggs   Wai-Yee Chan
Albert M. Chandler   A. Chadwick Cox
Jay S. Hanas   Robert E. Hurst
Leon Unger   Chi-Sun Wang

Springer-Verlag
New York Berlin Heidelberg London Paris
Tokyo Hong Kong Barcelona Budapest

ry and Molecular Biology
College of Medicine
Health Sciences Center
The University of Oklahoma
Oklahoma City, OK 73190
USA

Albert M. Chandler, Ph.D.
Department of Biochemistry and Molecular Biology
College of Medicine
Health Sciences Center
The University of Oklahoma
Oklahoma City, OK 73190
USA

Library of Congress Cataloging-in-Publication Data
Biochemistry / [edited by] Thomas Briggs, Albert M. Chandler : with
    contributions by Thomas Briggs . . . [et al.], — 2nd ed.
        p.   cm. — (Oklahoma notes)
      ISBN 0-387-97781-3. — ISBN 3-540-97781-3
      1. Biochemistry—Outlines, syllabi, etc.   2. Biochemistry—
Examinations, questions, etc.   I. Briggs, Thomas.   II. Chandler,
Albert M.   III. Series.
      [DNLM: 1. Biochemistry—examination questions.   2. Biochemistry—
outlines.   QU 18 B6154]
      QP518.3.B56   1992
      574.19′2—dc20
DNLM/DLC
for Library of Congress   92-2149

Printed on acid-free paper.

Production managed by Christin R. Ciresi; manufacturing supervised by Jacqui Ashri.
Camera-ready text prepared by the editors.
Printed and bound by Edwards Brothers, Inc., Ann Arbor, MI.
Printed in the United States of America.

9 8 7 6 5 4 3 2 1

ISBN 0-387-97781-3 Springer-Verlag New York Berlin Heidelberg
ISBN 3-540-97781-3 Springer-Verlag Berlin Heidelberg New York

Preface to the
Oklahoma Notes

In 1973, the University of Oklahoma College of Medicine instituted a requirement for passage of the Part 1 National Boards for promotion to the third year. To assist students in preparation for this examination, a two-week review of the basic sciences was added to the curriculum in 1975. Ten review texts were written by the faculty: four in anatomical sciences and one each in the other six basic sciences. Self-instructional quizzes were also developed by each discipline and administered during the review period.

The first year the course was instituted the Total Score performance on National Boards Part I increased 60 points, with the relative standing of the school changing from 56th to 9th in the nation. The performance of the class since then has remained near the national candidate mean (500) with a range of 467 to 537. This improvement in our own students' performance has been documented (Hyde et al: Performance on NBME Part I examination in relation to policies regarding use of test. J. Med. Educ. 60:439–443, 1985).

A questionnaire was administered to one of the classes after they had completed the Boards; 82% rated the review books as the most beneficial part of the course. These texts were subsequently rewritten and made available for use by all students of medicine who were preparing for comprehensive examinations in the Basic Medical Sciences. Since their introduction in 1987, over a quarter of a million copies have been sold. Assuming that 60,000 students have been first-time takers in the intervening five years, this equates to an average of four books per examinee.

Obviously these texts have proven to be of value. The main reason is that they present a *concise overview* of each discipline, emphasizing the content and concepts most appropriate to the task at hand, i.e., passage of a comprehensive examination over the Basic Medical Sciences.

The recent changes in the licensure examination structure that have been made to create a Step 1/Step 2 process have necessitated a complete revision of the Oklahoma Notes. This task was begun in the summer of 1991; the book you are now holding is a product of that revision. Besides bringing each book up to date, the authors have made every effort to make the texts and review questions conform to the new format of the National Board of Medical Examiners tests.

I hope you will find these review books valuable in your preparation for the licensure exams. Good Luck!

Richard M. Hyde, Ph.D.
Executive Editor

Preface to the Second Edition

As stated in the preface to the first edition, this book is intended to be a review and not a comprehensive textbook of Biochemistry and Molecular Biology. The book covers only the highlights from the much more detailed knowledge that is usually found in textbooks and we recommend that the reader turn to several excellent texts for more detailed reference. The book is intended to help those who are studying for National Medical Board Examinations and for similar examinations in the Allied Health fields. Although there is now a new form of unified medical examination, NMSLE Stage 1, the basic knowledge required to pass the biochemistry portion of this examination has not changed to any marked degree.

Two new chapters have been included that were not present in the first edition: Membranes and a chapter on Recombinant DNA Technology as related to medicine. The chapter on Genetic Diseases has been discontinued with the genetic information previously covered now dispersed through the individual chapters where appropriate. Each chapter has been carefully revised and rewritten where necessary and updated with the advent of new information. We have changed the question types that are included at the end of each chapter to conform more closely to the format now used in NMSLE examinations. As a general rule, words or other concepts which we consider to be of special importance have been portrayed in bold type, significant enzymes are in italics. Key sentences or concepts of special importance have been either italicized or underlined for emphasis.

The appearance of the book has been improved by printing it on a laser printer in proportional type and many of the drawings have been improved through the use of computer graphics. Errors in the first edition have been corrected and some difficult areas have been rewritten for greater clarity.

Each chapter has been contributed by a colleague who is an expert in the field or an experienced teacher of the subject. The words are the authors' own, but we have done some editing in order to achieve a reasonably consistent format. Because of the multiple authorship there is inevitably some unevenness in the depth of treatment of the various topics, but we accept responsibility for decisions on what to include and what to leave out. We welcome all comments and constructive suggestions for future editions.

Thomas Briggs
Albert M. Chandler

Contents

1. AMINO ACIDS AND PROTEINS

A. Chadwick Cox

I. AMINO ACIDS

A. The 20 Amino Acids. These are coded for in DNA:

Alanine	(Ala)	Leucine	(Leu)
Arginine	(Arg)	Lysine	(Lys)
Asparagine	(Asn)	Methionine	(Met)
Aspartic Acid	(Asp)	Phenylalanine	(Phe)
Cysteine	(Cys)	Proline	(Pro)
Glycine	(Gly)	Serine	(Ser)
Glutamine	(Gln)	Threonine	(Thr)
Glutamic Acid	(Glu)	Tyrosine	(Tyr)
Histidine	(His)	Tryptophan	(Trp)
Isoleucine	(Ile)	Valine	(Val)

B. Stereochemistry

The absolute configuration is the L (or S) enantiomer. The carbon adjacent to the carboxyl group is the α-carbon, to which the amino group is attached. Other carbon atoms are also attached to the α-carbon to form an R group. Amino acids differ from one another because they have different α-carbon R groups.

C. Classification: According to properties important to protein structure.

1. Hydrophobic (nonpolar): Ala, Cys, Gly, Ile, Leu, Met, Phe, Trp, Val.

2. Hydrophilic (polar and form H-bonds except for Pro)

    a. neutral: Asn, Gln, Pro, Ser, Thr, Tyr

    b. acidic (negative charge): Asp, Glu

    c. basic (positive charge): Arg, His, Lys

D. Post-translational Modifications

Several amino acids can be modified after being incorporated into proteins. Common post-translational modifications include formation of disulfide crosslinks, glycosylation of Ser or Asn residues, and phosphorylation of Ser, Thr and Tyr. The functions of these modification will be discussed later.

E. Amino Acids as Ampholytes

All the amino acids at neutrality possess both positive and negative charges and are called ampholytes and zwitterions. Because proteins contain mixtures of the acidic and basic amino acid residues indicated above, they are also ampholytes. Each of the ionizable groups on amino acids and proteins behaves according to the same principles that apply to any weak acid.

$$HA \rightleftharpoons H^+ + A^- \qquad\qquad K_a = [H^+][A^-] / [HA]$$

This equation can be written in a form that directly relates the concentrations to the pH — the Henderson-Hasselbalch equation:

$$pH = pK_a + \text{Log} ([\text{conjugate base}] / [\text{conjugate acid}])$$

When the conjugate base and acid are equal in concentration, their ratio is one, the log is zero, so pH = pK_a. The pH is buffered best by any ionizable group at or near its pK_a.

For carboxyl groups,

$$pH = pK_a + Log([COO^-]/[COOH])$$ pK_a usually in the 2 - 4 range

For amino groups,

$$pH = pK_a + Log([NH_2]/[NH_3^+])$$ pK_a usually in the 6 - 10 range

In histidine the N is part of an imidazole ring but it acts like an amino group with pK_a value around neutrality. For this reason His can participate in acid and base catalysis and is part of the catalytic mechanism of many enzymes.

As these equations indicate, the **charge ratio** varies with pH. An amino acid that has one amino group and one carboxyl group will be positively charged at low pH values. The net charge will diminish as the pH proceeds through values about the pK_a of the carboxyl group, because the latter are becoming negatively charged. It will equal zero at some pH intermediate between the pK values of the two ionizable groups, then will become negative as the pH proceeds through values about the pK_a of the amino group because these are losing their positive charge. The pH at which the net charge is zero is called the **isoelectric point** (pI) because the molecules in solution will not migrate in an electric field.

Figure 1-1 illustrates the points made above. The figure shows the number of equivalents of NaOH consumed by an amino acid in titrating the solution from pH 0 to pH 14. The amino acid has one basic and two acidic ionizable groups.

TITRATION OF AMINO ACID

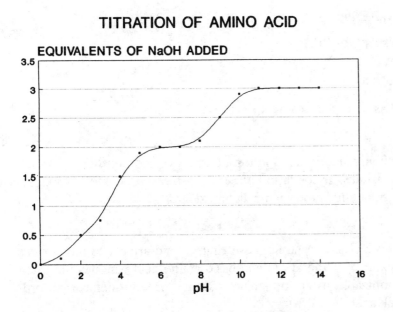

1. How many pK_a's are there? The same number as equivalents. (3)

2. What are their values? The same as the pH at each half equivalence. (About 2, 4, and 9 in this example).

3. Where is the isoelectric point? With two pK_a values in the acid region it must be an acid amino acid and the pI must be in the acid region halfway between the two pK_a values, that is, at pH 3.

Figure 1-1. Titration of an Amino Acid having Three Ionizable Groups.

II. PROTEIN STRUCTURE: GENERAL

A. The Peptide Bond

The peptide bond has partial double bond character because the double bond of the carbonyl group contributes electrons which allow different resonant forms to occur. Therefore, it is **planar** and only the *trans* configuration exists in proteins.

B. Steric Hindrance

The only rotation along the backbone of a polypeptide is the rotation of planes of the peptide bonds with respect to each other about the alpha carbons. However, this rotation is severely restricted by the steric hindrance between the different R groups and the carbonyl oxygens. One type of allowed rotation occurs when the carbonyl oxygens are pointed in the same directions and the R groups radiate out from the resulting spiral (as in the α helix). The other type occurs when alternate carbonyl oxygens point in opposite directions forming a pleated plane, with alternate R groups pointing up and down from this plane (as in β-pleated sheets).

C. Non-Covalent Bonds

The interactions between different parts of a protein are responsible for protein structure. These interactions are weak because they compete with interactions with water. They are:

Hydrogen bonds (H-bonds) between -C=O and -NH- of different peptide bonds; also involving some other amino acids capable of H-bonding.

Salt bonds (ionic) between acidic and basic amino acids.

Hydrophobic interactions among non-polar R groups.

Van der Waals forces.

D. Structural Levels of Proteins

Primary structure is the amino acid sequence within a peptide chain, including disulfide bonds and other covalent modifications.

Secondary structure consists of recognizable structures adopted by the backbone of the peptide segments. These are composed of the allowable sterically restricted types that also align H-bonds. That is, they are constrained by steric and H-bond requirements. The common secondary structures are:

1. right-handed α-helix
2. β-pleated sheet
3. β-turn

Tertiary structure is the specific three-dimensional conformation of a particular peptide chain. Most proteins are globular; hydrophobic interactions are primarily responsible for producing a compact conformation with a hydrophobic core.

Quaternary structure is the arrangement of subunits that are held together by non-covalent associations. An example is the tetrahedral arrangement of the four subunits of hemoglobin.

E. Domains

Most modern proteins are composites of evolutionarily older, smaller proteins. These smaller building blocks, called domains, often correspond to exons of the gene.

F. Conformational Changes

The tertiary structure of a protein changes when ligands such as substrates, effectors, cofactors, etc. bind to it. Changes in activity of an enzyme caused by changes in conformation are called **allosteric** effects.

G. Zymogens

Many enzymes are secreted as inactive precursors. These are either given the prefix "pro" as in pro-thrombin or the suffix "ogen" as in trypsinogen. Removal of small peptide fragment(s) by proteolysis activates the protein by causing a marked conformational change.

H. Isozymes

Often enzymes having multiple subunits but with the same catalytic function may differ from one another by slight variations in the primary sequence of one or more of the subunits. Lactic dehydrogenase is an example of an enzyme that occurs in different forms, called **isozymes**.

III. COLLAGEN AND FIBROUS PROTEINS

A. Collagen

Collagen is one of the most abundant of all proteins and is an important constituent of tendons, ligaments, cartilage, skin and the interstitial matrix.

Basic structure. Collagen is composed of three helical polypeptides all of which are wound tightly around each other to form a long, rigid fibrous molecule very unlike the usual globular proteins. The helical structures found in these polypeptides are not α helices because proline and hydroxyproline (~25% of the amino acids in collagen) do not fit the steric requirements of α helices.

Post-translational modifications

1. Two of the amino acids in collagen can be hydroxylated. These are **lysine** and **proline**. The formation of hydroxylysine and hydroxyproline increases the overall stability of the molecule. The hydroxylations are post-translational events and require the participation of **ascorbic acid** (Vitamin C). Ascorbic acid deficiency leads to the condition known as **scurvy** which is characterized by defective function of connective tissues and skin. (See Chapter 10). The hydroxylysine residues can also be glycosylated.

2. Lysine can also undergo oxidative deamination of its ϵ-amino group to form an aldehyde at this position. Two of these aldehyde residues on adjacent peptides can interact in an aldol condensation to form a covalent bond and increase the overall stability of the collagen molecule. Deficiency in the copper cofactor or in lysyl oxidase itself (Ehlers-Danlos syndrome) reduces the crosslinking in collagen and the desmosine crosslink in elastin. Another form of Ehlers-Danlos syndrome is caused by a deficiency in the propeptidase that converts procollagen to tropocollagen. This condition is characterized by stretchable skin and hypermobile joints.

Glycine is every third residue in a collagen peptide chain. Glycine is important because the degree of intertwining of the three chains is so tight that only glycine is small enough to fit at these positions.

Gene and exon structure. The collagen gene contains approximately 90 exons, the majority of which appear to have been constructed from an exon of (X-Y-Gly)$_6$.

B. Laminin and Fibronectin

These are fibrous proteins composed of multiple but dissimilar domains connected by flexible polypeptide segments. They are involved in cell-to-matrix interactions, each domain being specific in its binding to components of the extracellular matrix like collagen or to sites on cells.

IV. HEMOGLOBIN

A. Oxygen-Binding Proteins

Hemoglobin (Hb) is the oxygen-carrying system found in erythrocytes. It transports oxygen from the lungs to all tissues of the body, and aids in buffering blood.

Myoglobin (Mb) is a protein related to hemoglobin and is found in muscle. It can take oxygen from hemoglobin and acts as the intracellular oxygen-transport system.

Myoglobin	Hemoglobin
monomer	tetramer, 2α and 2β
binds one O_2	binds 4 O_2, one per subunit
binding property: hyperbolic	binding property: sigmoidal

B. Binding Curves (Figure 1-2)

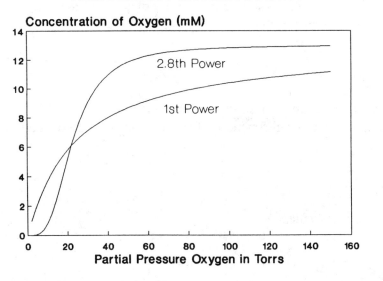

HEMOGLOBIN OXYGEN BINDING

The binding curve for hemoglobin measures the amount of O_2 bound by Hb *vs* the partial pressure of O_2. It is **sigmoidal** or S-shaped, indicating that hemoglobin is an allosteric and cooperative protein. The binding of O_2 to the first subunit causes a conformational change in the molecule which allows the other three O_2 molecules to bind much more readily. In the physiological range of oxygen concentrations, this sigmoidal binding curve is sharper (2.8th power) than the hyperbolic binding curve (1st power) exhibited by the monomeric and non-cooperative myoglobin. This sharpness allows more O_2 to be delivered for the same drop in oxygen tension between the lungs and other tissues.

Figure 1-2. Oxygen-Binding Curves of Hemoglobin and Myoglobin.

C. Regulation of O_2 Binding

The binding of O_2 to hemoglobin is regulated physiologically by three negative allosteric effectors: hydrogen ions, CO_2, and BPG. In the parlance of enzymologists, these effectors convert the conformation of hemoglobin from the "relaxed" to the "tense" state.

1. *Hydrogen ions*. When deoxygenated Hb binds O_2 in the lungs, hydrogen ions are released:

$$Hb(H^+)_2 + 4O_2 \rightleftharpoons Hb(O_2)_4 + 2H^+$$

In the tissues, metabolism releases acidic waste products (such as CO_2) causing an increase in hydrogen ion concentration. By mass action in the equation above this causes more O_2 to be released from the Hb. In the lungs CO_2 is excreted, the blood pH increases and Hb binds more O_2. This overall process, called the **Bohr effect**, has the result of increasing the amount of O_2 delivered to the more active tissues. At the same time more protons are removed from tissues and delivered to the lungs.

2. *Carbon dioxide* binds preferentially to deoxygenated Hb. It also has an allosteric effect like that produced by H^+. In the tissues both H^+ and carbon dioxide act additively to promote the release of oxygen.

3. *2,3-Bisphosphoglyceric acid* (BPG) also has an allosteric effect on Hb. BPG binds between the two β-subunits of deoxygenated Hb (one BPG per Hb) and stabilizes the molecule. BPG causes the steepest part of the oxygen binding curve to occur at an oxygen tension between that in the peripheral tissues and that in the lung. This arrangement permits the greatest amount of oxygen to be delivered to the tissues.

D. Sickle Cell Anemia

A reduction in the amount of hemoglobin in blood is referred to as anemia and of course leads to a reduction in the amount of O_2 carried. Sickle cell anemia is a heritable type resulting from a single amino acid mutation of a Val for a Glu in the β subunit (Hb S). While this has no effect on O_2 binding, Hb S in the deoxygenated state precipitates and forms fibers that distort the erythrocytes, which are then destroyed leading to anemia. Although the homozygous state is detrimental to survival, heterozygotes possess increased resistance to malaria. Hemoglobin S thus has net survival value in some geographic areas.

V. PLASMA PROTEINS

Plasma contains many proteins of a non-enzymatic nature. These include the immunoglobulins, blood coagulation proteins, transport proteins and proteins of unknown function. The function of some of these proteins are described in other sections.

A. Transport or Carrier Proteins

Functions

-Increase the water-solubility of hydrophobic molecules.

-Decrease the loss of small molecules in the kidney.

-Transport bound molecules to specific tissues.

-Assist in detoxification.

Selected examples:

Albumin makes up 50 to 55% of the proteins of plasma, and has very broad and non-specific binding properties. It binds and transports fatty acids released from adipose tissue, but also binds many drugs which compete with fatty acid binding. It does not bind steroids. Albumin is the main contributor to the osmotic pressure of blood. Severe loss of albumin causes edema and disturbances in blood volume and pressure.

Transferrin carries two Fe^{+++} ions and transfers them to cells having receptors for transferrin. **Ferritin** is the intracellular form of iron storage, but small amounts also occur in blood.

Haptoglobin and **hemopexin** bind methemoglobin and hemin, respectively. Binding hemin not only protects the kidneys from its toxic effects but also conserves the iron for re-use in metabolism.

B. Trace Enzymes

In trace amounts, blood contains many enzymes that come from dying cells. An increased level of an enzyme is diagnostic for trauma to those tissues that are rich in the specific enzyme. Some important examples are listed under the enzyme section of this text.

VI. HEMOSTASIS AND BLOOD COAGULATION

A. The Hemostatic Process

Hemostasis, the stopping of blood flow at a wound site, is brought about by the combined effect of platelets, the vessel wall, and plasma coagulation factors. The chronology of hemostasis is that a wound exposes the collagen layer just below the endothelial cells that line the vessel wall. Platelets recognize and bind to the collagen, become activated, recruit other platelets to form a hemostatic plug, secrete many substances that promote coagulation, and support coagulation once it starts.

Cast of characters. Table 1-1 lists the hemostatic and fibrinolytic factors, the helper proteins, and the protease inhibitors that control the proteolytic factors.

Factor	Trivial Name	Helper Protein	Inhibitor in Plasma
I	Fibrinogen		
II	Prothrombin, Thrombin	Thrombomodulin	Antithrombin III
III	Tissue Factor		
IV	Calcium ions		
V	Plasma accelerator globulin		Protein C
VI	(Not Assigned)		
VII	Proconvertin	Tissue Factor	Antithrombin III
VIII	Antihemophilic factor A		
IX	Antihemophilic factor B	F VIII	Antithrombin III
X	Stuart factor	F V	Antithrombin III
XI	Thromboplastin antecedent		
XII	Contact factor, Hageman		
XIII	Transglutaminase		
	Tissue plasminogen activator	Fibrin clot	Inhibitors of plasminogen activator
	Plasminogen/plasmin		Anti$_2$-plasmin
	Protein C	Protein S	Protein C inhibitor
	Protein S		

Table 1-1. Hemostatic and Fibrinolytic Factors, their Helper Proteins, and Inhibitors.

The plasma coagulation proteins are numbered by Roman numerals and a subscripted "a" indicates they have been converted into their active forms. Prothrombin (Factor II, or F II), thrombin (F II$_a$, the active form of prothrombin), and fibrinogen (F I), are almost always still referred to by their trivial names. Plasma also contains many protease inhibitors some of which control coagulation.

Wound recognition and formation of platelet plug. When a wound occurs, the thin layer of endothelial cells lining blood vessels is disrupted, exposing collagen in the basement membrane below. Platelets bind to this collagen and become active. To build the plug, activated platelets recruit other platelets by secreting the platelet agonists ADP, serotonin, and thromboxane A$_2$. The latter two help hemostasis by causing smooth muscle cells of vessel walls to contract. The aggregated platelets occlude the vessels but do not provide a sufficiently stable plug to prevent re-bleeding. Coagulation of the blood provides the added stability.

Initiation of coagulation. In addition to exposing collagen, the wound also exposes cells that have a receptor, called tissue factor, that binds F VII and F VII$_a$. In Figure 1-3, adventitial cells are shown supplying tissue factor but other cells provide it as well. Exposure of tissue factor starts coagulation.

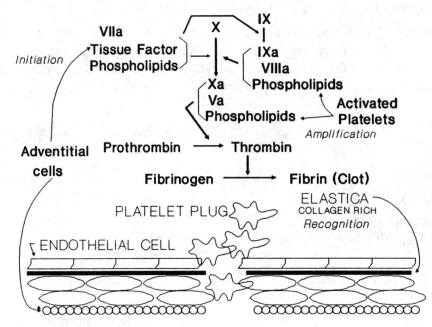

Figure 1-3. The Coagulation Cascade.

The cascade of zymogen activation — amplification: When F VII$_a$ is bound to tissue factor, which is also a helper protein, it proteolytically activates F X and to a lesser extent F IX. F IX$_a$, with its helper protein F VIII$_a$, also activates F X. In turn, F X$_a$, with its helper F V$_a$ activates prothrombin to thrombin. Because this cascade is composed of a series of enzymes activating other enzymes, it takes only a small amount of F VII$_a$ bound to tissue factor to produce large quantities of thrombin. Furthermore, F X activates F VII on tissue factor and thrombin activates F V and F VIII, each to promote its own production.

This scheme also provides a degree of redundancy. That is, the step involving Factors IX and VIII is only supportive in many tissues. Classical hemophilia results from production of a mutant, inactive F VIII. Bleeding in tissues rich in tissue factor during mild trauma is not a major problem for hemophiliacs because of the direct activation of F X. Hemophiliacs suffer mainly chronic bleeding into joints and muscles where cells rich in tissue factor are sparsely distributed.

The activations are all proteolytic cleavages, each requiring the following agents:

The *activating factor*, in each case a serine protease: $F VII_a$, $F IX_a$, and $F X_a$.

Substrate: F X, F X, and prothrombin, respectively.

Helper protein: tissue factor, $F VIII_a$, and $F V_a$ respectively. A helper protein aids in holding the substrate in place during the activation (it lowers the apparent K_M). Tissue factor is an integral protein of the membranes of certain cells. Factors V_a and VIII_a are not integral proteins but are tightly bound to platelet membranes.

Acidic phospholipid surfaces, needed for sufficient reaction velocities.

γ-Carboxy-glutamyl residues, whose post-translational production is vitamin K-dependent.

Calcium ions which, through the γ-carboxy-glutamyl residues, form bridges between the proteases and phospholipid surfaces, and between the latter and the substrates of the proteases.

Factors II, VII, IX, X, and proteins C and S (to be discussed below) share the requirement for the modified glutamyl groups because binding to phospholipids is an integral part of their function. Warfarin, a competitive inhibitor of vitamin K - dependent processing, and vitamin K deficiency prevent the formation of the γ-carboxy-glutamyl residues and thereby reduce hemostasis.

Coagulation is restricted to the wound site because the activating reactions occur only when damaged tissue binds the activated platelets that then supply acidic phospholipid surfaces for the proteolytic steps catalyzed by $F IX_a$ and $F X_a$. While bound with their helper proteins to the platelets, these proteases are also protected from the inhibitors of plasma proteases, antithrombin III and α-macroglobulin. Furthermore, the activation reactions occur too slowly if the activating factors are not bound to activated platelets.

Clot formation. The clot forms as fibrin is produced between the aggregated platelets. Thrombin is the first protease that is freed from the platelets in a viable, active form. It causes the clot by cleaving fibrinogen, removing fibrinopeptides A and B. Fibrin, the product of the cleavage, polymerizes on its own. As with other fibrous proteins, fibrin polymers are cross-linked by $F XIII_a$, a transglutaminase that joins the side chains of a Gln on one fibrin molecule to a Lys on a different fibrin molecule. Thrombin also activates F XIII. Platelets aggregate by binding to either end of fibrinogen which then forms a bridge between two platelets. Fibrin strands still have these same ends, therefore platelets bind fibrin, help in organizing the fibrin strands and even pull these into a tighter form to maintain a strong clot. These fibrin strands suture the platelets together to provide the needed stability that the platelet plug alone does not possess.

Summary of the hemostatic process. There are two central elements in coagulation. One is the platelet that recognizes the wound, provides the surface for activation steps, secretes many substances that affect coagulation, and organizes and retracts the clot. The other is thrombin which causes fibrinogen to clot, further activates platelets, stimulates its own production by activating Factors V and VIII, activates Factor XIII, and even initiates the reactions that regulate its own production.

B. Regulation of Hemostasis

Obviously, from the description above something is required to prevent the thrombin and unincorporated, activated platelets from spreading the clot throughout the body. This something is a separate collection of reactions primarily under the control of the monolayer of endothelial cells that line the blood vessels.

Termination of coagulation. Any thrombin that escapes from the area of the clot is rapidly bound to the surface of intact endothelial cells whereupon it activates the anticoagulation scheme shown in Figure 1-4. Thrombin binds to thrombomodulin, its receptor on endothelial cells, which also binds to protein C. Thrombomodulin acts like a helper protein in the activation by thrombin of protein C, which in turn with its helper protein, protein S, inhibits $F V_a$ and $F VIII_a$ on activated platelets. This is the reaction mechanism that ensures that even if an activated platelet escapes the hemostatic plug, the coagulation process will be terminated. Eventually these factors terminate the clotting reactions at the wound site as well.

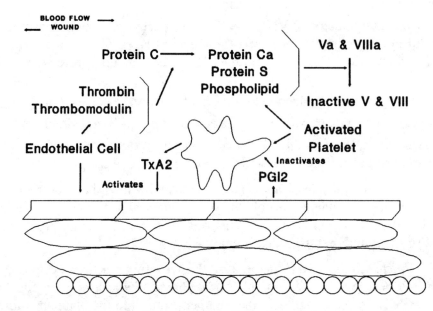

Figure 1-4. Regulation of Coagulation Downstream from Wound.

Inactivation of activated platelets. Endothelial cells also deactivate activated platelets. Thrombin and thromboxane both stimulate endothelial cells to produce prostacyclin I_2 (PGI$_2$). Prostacyclin I_2 is a potent antagonist that deactivates platelets. It also is a potent vasodilator that overcomes the contraction of the vessel wall induced by the platelet-derived serotonin and thromboxane A_2.

C. Fibrinolysis or Removal of the Clot

The clot is removed as part of the process of wound healing (Figure 1-5). Fibrin clots are removed proteolytically by plasmin. Plasmin is activated on the clots by plasminogen activator, a protein released from endothelial cells in response to the clotting process. The fibrin clot acts as a helper protein for this activation. Both plasminogen, the inactive form of plasmin, and the activator are bound to the clot and the activation is enhanced by the clot. This leaves plasmin on the clot until it finishes the hydrolysis. Antiplasmin, a protein that inhibits plasmin, is present in plasma to inhibit any plasmin that escapes from the clot prematurely. In this way, clot removal is also restricted to the wound site.

D. Thrombosis

Thrombosis is abnormal coagulation that occurs on atherosclerotic plaques and contributes to a variety of vascular problems including myocardial infarctions, strokes and deep-vein thrombosis. Several methods of controlling this disorder have been tried, most based on the material presented above. Another method is based on reducing risk factors. This approach depends primarily on educating the public as to the benefits of diet and exercise in decreasing the severity of atherosclerosis that comes with age.

Regulating platelet function. Attempts to control thrombosis by regulating platelet function have been of limited value. **Aspirin** has received the most attention but it appears to provide limited aid in preventing strokes or recurrent heart attacks. Aspirin is an irreversible inhibitor of *cyclooxygenase*, an enzyme necessary for the synthesis of thromboxane A_2, the "message for help" sent out by activated platelets. One problem with aspirin is that it blocks cyclooxygenase in endothelial cells as well and thereby decreases the production of the natural anticoagulant, prostacyclin I_2.

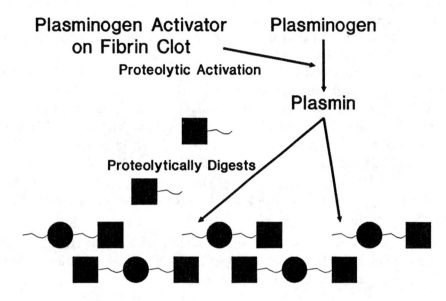

Figure 1-5. Fibrinolysis (clot digestion).

Regulating coagulation. **Warfarin** and other coumarin drugs have been used in the past for prophylaxis against thrombosis. Recall that warfarin prevents the post-translational modification necessary for binding of some coagulation factors to the reactive phospholipid surfaces of platelets. However, the efficacy of coumarin drugs was poor and the discovery of the regulatory role of proteins C and S, both of which contain the vitamin K-dependant modifications, explains the deficiency in effect. But since these drugs can be administered orally, and do have some prophylactic effect, they are useful for certain types of patient care.

Heparin, an anticoagulant, also is used to control thrombosis. (Chelators of calcium ions cannot be used because the calcium ion concentration in blood is very critical for many processes). Heparin acts as a catalyst for the inactivation of thrombin by antithrombin III, the plasma protease inhibitor. Heparin is one of the favored agents for regulating deep vein thrombosis in many types of patients and in preventing recurrent clot formation after clot removal. Heparin must be injected for this treatment and is not suitable as a long-term prophylactic.

Increasing fibrinolysis. Another approach used to treat thrombosis is to activate the fibrinolysis reactions. **Plasminogen activator**, *urokinase*, and *streptokinase*, all activators of plasminogen, have been used in treating thrombosis on an acute basis. Streptokinase, a bacterial product, was the only one inexpensive enough for routine use until all were cloned into bacteria or culture cells. At first, recombinant tissue plasminogen activator (TPA) was preferred because it was thought that there would be reduced problems of bleeding since it requires a clot for activation. But several studies suggest that there are few differences among fibrinolytic agents in removing clots. However, the foreign nature of streptokinase can add complications for patients previously exposed to this agent.

E. Coagulation Deficiencies

Thrombosis is the most common problem associated with the clotting system but there are examples of deficiencies in all the factors (see Table 1-2, page 13). There are also various platelet disorders and defects in vessel wall reactions. The best known example is classical **hemophilia**, which results from deficient activity of F VIII. These conditions are most often treated by replacement therapy, in which major complications can arise from blood-borne diseases and immune responses.

Clinical Tests. Coagulation can be assayed readily in the clinical laboratory using plasma samples prepared from anticoagulated blood. These tests are useful in screening for possible coagulation deficiencies, the nature of which can be established later by more specific tests. **Prothrombin time** is a test that measures the time required for the plasma to clot after the reaction is initiated by addition of thromboplastin, a mixture of tissue factor and phospholipids. This test bypasses the step that requires a complex of F IX and its helper F VIII. Because tissue factor was added and this is not a plasma protein, this pathway is called the **extrinsic pathway**. In contrast, the **partial thromboplastin time** is a test which measures the time required for a clot to form after the reaction is initiated by addition of phospholipid and a contact activator. The contact activator activates F XII which activates F XI, which in turn activates F IX. The rest of this pathway, called the **intrinsic pathway** because no non-plasma factors were added, is the same as the process discussed above. The physiological relevance of this pathway of activation of F XII is not understood but the diagnostic virtue is. Thus hemophilia A (F VIII deficiency) or B (F IX deficiency) prolong the partial thromboplastin time (the longer pathway with the longer-named test) but not the prothrombin time. Warfarin and the coumarins prolong both times.

Platelet function is tested by the **bleeding time**. Cessation of bleeding from a standardized cut in the skin depends on formation of a plug by platelets. Aspirin, which blocks production of thromboxane A_2, prolongs bleeding time, whereas hemophilias and warfarin do not.

VII. GENETIC DISEASES

A. Inherited Structural Variants of Hemoglobin

A single base change (point mutation) may occur in the gene for the α, β, γ, or δ-globin chain. This may cause a normal (non-deleterious) amino acid substitution, eg., Hb C $\beta^{6\ Glu\ \to\ Lys}$; Hb E $\beta^{26\ Glu\ \to\ Lys}$, or it may result in production of an abnormal protein. It may also be a nonsense mutation, causing chain termination. Deletion (or insertion) of a single base causes a frameshift mutation. Mutant proteins occur in which amino acids are deleted or inserted, resulting from deletion or insertion of entire codons.

The defective molecule in **sickle cell anemia** is Hb S, which has a single amino acid substitution in the β chain — 6 Glu → 6 Val (see page 6).

B. Quantitative Disorders of Globin Synthesis: The Thalassemias

These diseases are characterized by abnormalities in the amounts of the different globin chains synthesized. The chains themselves are normal in structure.

α-*Thalassemias*: deficiency of α chain synthesis.

-homozygous for α thalassemia: all 4 α globin genes — Hb Barts, death in utero

-heterozygous for α thalassemia: 3 globin genes affected — Hb H disease

-α_1-thalassemia: 2 α globin genes affected

-α_2-thalassemia: 1 α globin gene affected.

β-*Thalassemias*: decreased (β^+) or absent (β^0) synthesis of β-chain results in increased levels of Hb F and Hb H2.

-homozygous condition: thalassemia major — Cooley's anemia, severe microcytic anemia.

-heterozygous condition: thalassemia minor.

C. Collagen metabolism

Ehlers-Danlos syndrome, types I - VII. As a group, these diseases are characterized by deficiencies at various stages in the formation of collagen, from synthesis of protein to cross-linking of chains.

Osteogenesis imperfecta, types I - III: defective synthesis or secretion of types I and/or III collagen.

Cutis laxa. An X-linked recessive defect causing deficient lysyl oxidase with abnormal copper metabolism.

D. Blood coagulation: Deficiency of Clotting Factors

Disorder	Factor	Genetics
Afibrinogenemia	Fibrinogen	Autosomal recessive
Dysfibrinogenemia	Fibrinogen	Autosomal recessive
Parahemophilia	V	Autosomal recessive
Factor VII deficiency	VII	Autosomal recessive
Classic hemophilia	VIII	X-linked recessive
Hemophilia B	IX	X-linked recessive
Factor X deficiency	X	Autosomal recessive
Factor XI deficiency	XI	Autosomal recessive
von Willebrand	von Willebrand	Mostly autosomal dominant

Table 1-2. Heritable Diseases of Blood Coagulation.

VIII. REVIEW QUESTIONS ON AMINO ACIDS AND PROTEINS

DIRECTIONS: For each of the following multiple-choice questions (1 - 34), choose the ONE BEST answer.

The figure below refers to Questions 1 and 2.

The figure below refers to Questions 3 and 4:

HEMOGLOBIN OXYGEN BINDING

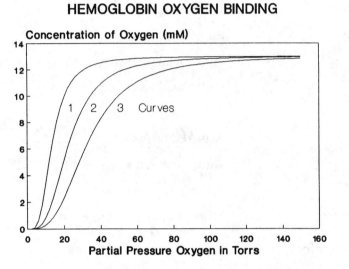

TITRATION OF AMINO ACID
Equiv. of NaOH consumed by Amino Acid

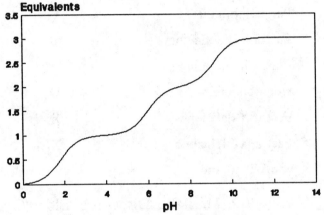

1. Curve 2 represents the O_2 binding curve for hemoglobin. An increase in the CO_2 concentration without a change in pH would cause the curve to:

A. shift to curve 1
B. remain the same
C. remain in the same position as curve 2 but decrease in magnitude
D. shift to curve 3
E. none of the above.

2. If the partial pressure of O_2 remains the same, what effect would anemia have on the binding curve? (Note the units of concentration).

A. shift to curve 1
B. remain the same
C. remain in the same position as curve 2 but decrease in magnitude
D. shift to curve 3
E. none of the above.

3. Which one of the following amino acids can fit the curve shown?

A. Ser
B. Asp
C. Pro
D. His
E. Gly.

4. What is the net charge on the molecule at pH 6?

A. one net negative charge
B. half the molecules have one net negative charge and half have a net charge of zero
C. half the molecules have one net positive charge and half have a net charge of zero
D. half the molecules have two net positive charges and half have one net positive charge
E. two net positive charges.

5. Therapy of thrombosis with warfarin has proven less effective than originally expected because:

A. vitamin K is too plentiful in our diets
B. of warfarin's severe side effects
C. warfarin also reduces the synthesis of functional proteins C and S
D. no adequate clinical test exists to monitor the dose
E. all of the above.

6. One role of a helper protein is to:

A. provide the active site for the complex between a protease and its substrate
B. transport a small molecule and insure that it is delivered to the targeted tissue
C. decrease the apparent K_M of the reaction involving a protease
D. transport small molecules across cellular membranes
E. none of the above.

7. The Henderson-Hasselbalch equation can NOT be used to calculate the:

A. pH of a solution if the pK_a and the ratio of conjugate acid to conjugate base are known
B. pK_a if the hydrogen ion, conjugate base form and the total weak acid concentrations are known
C. ratio of charged to the uncharged form if the K_a and hydrogen ion concentration are known
D. pK_a when the concentration of the conjugate base is exactly equal to that of the conjugate acid
E. concentration of conjugate base when a known amount of the conjugate acid was added to a solution of conjugate base to adjust the pH = pK_a.

8. Hemophilia, caused by a deficiency in Factor VIII, is characterized by an increase in:

A. bleeding time
B. partial thromboplastin time
C. prothrombin time
D. B and C are correct
E. A, B and C are correct.

9. The differences in the binding curves between hemoglobin and myoglobin are due primarily to:

A. the amount of oxygen that can be carried by the different heme groups
B. the markedly different tertiary structures
C. interactions among subunits
D. the differences in pH of the surrounding solutions
E. substitution of a lysyl for a histidyl group.

10. The interaction of hemoglobin with oxygen:

A. is reversible
B. has the stoichiometry of one O_2 bound per each heme prosthetic group
C. displays a 2.8 power of the concentration in the saturation curve
D. decreases the binding of BPG (bisphosphoglycerate) to hemoglobin
E. all of the above.

11. Coagulation is restricted to the wound site because:

A. fibrinogen is selectively secreted at the wound site
B. the formation of plasmin is prevented at the clot site
C. activation of factor X and prothrombin occurs only on activated platelets
D. the stabilization of the clot occurs exclusively at the wound site where activated endothelial cells secrete Factor XIII_a
E. exposure to air is required.

12. The pleated sheet structure (β-conformation):

A. contains parallel chains which are stabilized only by hydrophobic interactions between side chains of the amino acids
B. is pleated because the peptide bond is not planar
C. is found in collagen
D. has the side chains of adjacent amino acids alternating above and below the plane of the sheet
E. can only be formed in the test tube.

13. The occurrence of a wound produces a signal that is recognized by platelets. This signal is:

A. generation of thrombin
B. generation of plasmin
C. production of prostacyclin I_2
D. exposed collagen
E. cyclic AMP.

14. Albumin is all of the following EXCEPT:

A. the major carrier of sterols
B. a transporter of fatty acids
C. the common drug binding protein
D. the major protein contributor to the osmotic pressure of plasma
E. the major plasma protein.

15. The endothelial cell may be considered the principal anticoagulant cell because it performs all the following activities EXCEPT:

A. produces prostacyclin I_2
B. secretes plasminogen activator
C. supports protein C activation
D. consumes protein S
E. aids in the inactivation of thrombin.

16. Many proteins contain two or more segments (domains) displaying homology in sequence. These proteins have evolved by:

A. allosteric cooperativity
B. development of disulfide crosslinking
C. gene duplication and fusion
D. post-translational modification
E. transpeptide formation.

17. Ascorbic acid (vitamin C), needed for the synthesis of functional collagen, is required for reactions catalyzed by:

A. lysine hydroxylase
B. lysine oxidase
C. proline hydroxylase
D. proline oxidase
E. Both A and C are correct.

18. The distortion in the shape of red blood cells caused by sickle cell hemoglobin (HbS) is due to:

A. an abnormal porphyrin ring structure
B. decreased binding capacity for oxygen
C. abnormal solubility of the deoxy form of the protein
D. a shortened α chain
E. inability to bind the heme group efficiently.

19. Which of the bonds listed below is NOT an example of a covalent crosslink found in a fibrous protein?

A. desmosine
B. lysyl aldol bond
C. glutaminyl-lysine transpeptide bond
D. salt bridge
E. disulfide bond.

20. In which of the following groups do all of the amino acids have basic side chains?

A. Lys, His and Phe
B. Lys, His and Arg
C. Trp, Lys and Arg
D. Lys, Arg, Gln and Asn
E. Trp, Tyr, His and Arg.

21. Hemoglobin is a good example of a protein having:

A. allostericity
B. cooperativity
C. multiple subunits
D. quaternary structure
E. all of the above.

22. The three individual polypeptide helices of collagen are tightly wound about each other in a super helix. This winding requires:

A. hydroxylated lysines
B. oxidized lysines
C. hydroxylated prolines
D. periodicity in the positions of glycine
E. all of the above.

23. Most of the amino acids in the core of a protein are classified as:

A. conservative
B. divergent
C. helical
D. hydrophobic
E. reversible.

24. Thrombin, the first free protease produced by the coagulation cascade, activates all of the following EXCEPT:

A. Factor V
B. Factor VIII
C. Factor XI
D. Factor XIII
E. protein C.

25. The common interactions responsible for tertiary structure are:

A. hydrogen bonds
B. ionic bonds
C. hydrophobic interactions
D. steric hindrance
E. all of the above.

26. Clot digestion is restricted to the clot for all the following reasons EXCEPT:

A. plasminogen/plasmin binds to the clot
B. plasminogen is activated on the clot
C. tissue plasminogen activator binds to the clot
D. activated platelets are required for plasminogen activation
E. antiplasmin inactivates unbound plasmin.

27. Active helper proteins are inactivated downstream from the wound site by:

A. α_2-macroglobulin
B. antiplasmin
C. activated protein C
D. heparin
E. thrombomodulin.

28. Activation of a zymogen is accomplished by:

A. selective coupling of an activator peptide to an inactive precursor
B. selective cleavage of specific bonds in an inactive precursor
C. selective removal of coenzyme molecules
D. selective addition of prosthetic groups
E. all of the above.

29. Plasma transport proteins perform all of the following functions EXCEPT:

A. reduce loss of molecules through the kidney
B. render molecules less toxic
C. serve as specific markers of tissue damage
D. solubilize hydrophobic molecules
E. target molecules to specific tissues.

30. The effect of a normal dose of aspirin would be to increase:

A. bleeding time
B. partial thromboplastin time
C. prothrombin time
D. A and B are correct
E. B and C are correct.

31. Which one of the following pairs does NOT represent an inhibitor and what it inhibits?

A. prostacyclin I_2 – activation of platelets
B. thromboxane A_2 – activation of endothelial cells
C. protein C & protein S – Factor VIII_a
D. antithrombin III – thrombin
E. antithrombin III – Factor X.

32. Increasing the partial pressure of O_2 (P_{O_2}) of venous blood would cause (in the venous blood):

A. increases in pH, P_{CO_2} and concentration of free BPG (bisphosphoglycerate)
B. increases in P_{CO_2} and free BPG
C. a decrease in free BPG
D. a decrease in P_{CO_2} and free BPG
E. a decrease in pH, P_{CO_2} and free BPG.

33. Thrombomodulin:

A. is bound to endothelial cells
B. functions as an anticoagulant
C. is a binding site for thrombin
D. is a binding site for protein C
E. all of the above.

34. The peptide bond:

A. is planar
B. has resonant double bond character
C. is *trans* in configuration
D. rotates about the bond to the alpha carbon
E. all of the above.

IX. ANSWERS TO QUESTIONS ON AMINO ACIDS AND PROTEINS

1.	D	13.	D	25.	E
2.	C	14.	A	26.	D
3.	D	15.	D	27.	C
4.	C	16.	C	28.	B
5.	C	17.	E	29.	C
6.	C	18.	C	30.	A
7.	D	19.	D	31.	B
8.	B	20.	B	32.	B
9.	C	21.	E	33.	E
10.	E	22.	D	34.	E
11.	C	23.	D		
12.	D	24.	C		

2. ENZYMES

Wai-Yee Chan

I. NATURE OF ENZYMES

A. Introduction

All known enzymes are proteins. One exception to this statement might be the self-splicing of ribosomal RNA which does not involve any protein molecule. RNA serves as enzyme in this case.

Enzymes are **biological catalysts**, produced by living tissues, that increase the **rates** of reactions. They do not affect the nature of an equilibrium; they merely speed up the rate at which it is achieved.

B. Definition of Terms

Substrate: substance acted upon by an enzyme.

Activity: amount of substrate converted to product by the enzyme per unit time (e.g. micromoles/minute).

Specific activity: activity per quantity of protein (e.g. micromoles/minute/mg protein).

Catalytic constant: proportionality constant between the reaction velocity and the concentration of enzyme catalyzing the reaction. Unit: activity/mole enzyme.

Turnover number: catalytic constant/number of active sites/mole enzyme.

International Unit (IU): quantity of enzyme needed to transform 1.0 micromole of substrate to product per minute at 30°C and optimal pH.

C. Nomenclature

Some enzymes have trivial names, e.g. *pepsin*, *trypsin*, etc. Others are named by adding the suffix -ase to the name of the substrate, e.g. *arginase*, which catalyzes the conversion of arginine to ornithine and urea. All enzymes have systematic names, of which there are 6 major classes:

1. *Oxidoreductase*: oxidation-reduction reactions, e.g., *alcohol: NAD oxidoreductase* for the enzyme catalyzing the reaction:

$$RCH_2OH + NAD^+ \rightleftharpoons RCHO + NADH + H^+.$$

2. *Transferase*: transfer of functional groups including amino, acyl, phosphate, one-carbon, glycosyl, etc. Example: *ATP: Creatine phosphotransferase* for the enzyme catalyzing the reaction:

$$ATP + Creatine \rightleftharpoons Phosphocreatine + ADP.$$

3. *Hydrolase*: Cleavage of bond between carbon and some other atoms by the addition of water. Example: *Peptidase* for the enzyme catalyzing:

$$R_1CONHR_2 + H_2O \rightleftharpoons R_1COOH + R_2NH_2.$$

4. *Lyase*: Addition or removal of the elements of water, ammonia or CO_2 to or from a double bond. Example: *Phenylalanine ammonia lyase* for the enzyme catalyzing the reaction:

$$phenylalanine \rightleftharpoons cinnamic\ acid + ammonia.$$

5. *Isomerase*: Racemization of optical or geometric isomers. Two types:

    a. *epimerase* or *racemase* for optical isomers (asymmetric carbon), as in:

        D-lactic acid $\rightleftharpoons$ L-lactic acid (*racemase*)

    b. *mutase* for geometric isomers or intramolecular group transfer, as in:

        2-phosphoglycerate $\rightleftharpoons$ 3-phosphoglycerate

6. *Ligase*: Formation of C-O, C-S, C-N, or C-C with the hydrolysis of ATP. Example: *Pyruvate carboxylase* for the enzyme catalyzing the reaction:

        pyruvate + ATP + CO_2 $\rightleftharpoons$ oxaloacetate + ADP + P_i

D. Basic Enzyme Structure

Enzymes may be composed of a **single** polypeptide chain, or several identical or different **subunits**. Different enzymatic activities may be contributed by independent **domains** of a single polypeptide.

A compound (organic or inorganic) other than an amino acid side chain may be required for activity and is not modified at the end of the reaction When tightly bound to an enzyme, such as the heme of cytochrome, it is known as a **prosthetic group**. When less tightly bound, or removable by dialysis, e.g. metal ions, it is known as a **cofactor**. The catalytically active enzyme complex consisting of a protein **apoenzyme** and a non-protein cofactor is called a **holoenzyme**.

Coenzymes are organic molecules fulfilling the role of substrate, being modified at the end of the reaction, but readily regenerated by another linked reaction. Examples: biotin, NAD, ATP, TPP, FAD, pyridoxal phosphate, coenzyme A, etc.

When an enzyme is first made as an activatable precursor, it is called a **zymogen**, or **proenzyme**. The active mature enzyme is generated by specific cleavage of a peptide bond. Examples:

        chymotrypsinogen → *chymotrypsin*;   pepsinogen → *pepsin*

E. Characteristics of Enzymatic Reactions

Enormous catalytic power.

Optimum pH (Figure 2-1): the pH at which the enzyme activity is at its maximum. This depends on the acid-base behavior of the enzyme and substrate. It can vary widely; the majority of enzymes have optima between 4 and 8. The kinetic parameters K_M and V_{max} vary independently at different pH's. At extreme pH or temperature, enzymes are **denatured**.

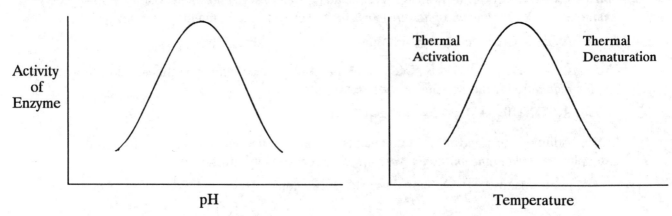

Figure 2-1. Enzyme Activity as a Function of pH and Temperature.

Optimum temperature (Figure 2-1): the temperature at which the enzyme activity is at its maximum. This varies according to conditions such as salt content, pH, etc. But enzymes from different tissues of the same organism do not necessarily have the same temperature profile. The rate of most enzymatic reactions about doubles for each $10°C$ rise in temperature.

Saturatability. A maximum velocity is not exceeded even in the presence of excess substrate.

Reaction velocity is directly proportional to enzyme concentration, provided there is enough substrate. The enzymatic reaction continues until substrate is exhausted.

Substrate specificity can be high or even absolute (one particular compound). Specificity can apply broadly to a class of compounds sharing a type of linkage, steric structure (cis-trans), or optical activity (D,L).

Regulation of activity is by **feedback inhibition**, altered **availability of substrate**, or altered **kinetic parameters**.

II. ENZYME KINETICS

A. Basic Principles

Order of reaction. If $A \rightleftharpoons P$, then $v = k[A]^R$, where

    v = velocity of reaction

    k = rate constant

    R = order of reaction

    [A] = concentration of reactant.

Energetics of catalysis (Figure 2-2). The rate of the reaction depends on the number of activated molecules in the **transition state** (activated complex A - - - B). The **free energy of activation** (E_A, or $\Delta G^{\ddagger}$ or $\Delta F^{\ddagger}$) is the amount of energy which must be put into the system to reach the activated transition state. A catalyst forms a transition complex with a **lower E_A(cat)**. Since an enzyme lowers the E_A for both the forward and back reactions, the velocity for both reactions is faster and equilibrium is achieved sooner.

The **free energy change** (ΔG or ΔF) of the reaction is the difference in free energy between reactants and products. It is NOT changed in the presence of a catalyst.

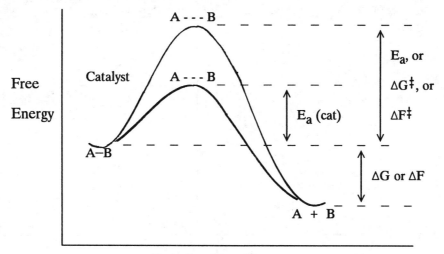

Figure 2-2. Energetics of Enzyme Catalysis.

Equilibrium constant — equilibrium concentrations of products multiplied, divided by equilibrium concentrations of reactants multiplied. Its value is NOT changed in the presence of a catalyst:

$$K_{eq} = \frac{[P_1] \cdot [P_2] \cdot \ldots}{[A_1] \cdot [A_2] \cdot \ldots}$$

B. Kinetics

Michaelis-Menten Equation

Given the reaction:

$$E + S \underset{k_2}{\overset{k_1}{\rightleftharpoons}} ES \xrightarrow{k_3} E + P$$

then $v = \dfrac{V_{max}(S)}{K_M + (S)}$

where:

E = enzyme
S = substrate
ES = enzyme-substrate complex
P = product
k_1,k_2,k_3 = rate constants

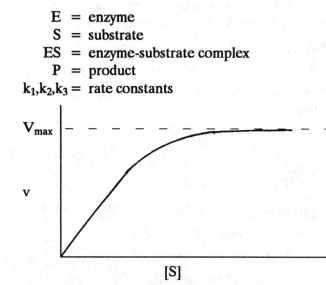

v (Figure 2-3) is the **initial velocity** of the reaction at time essentially 0, when there is no P. Therefore the back reaction ES ← E + P can be disregarded.

V_{max} is the **maximal velocity** achieved when the enzyme is saturated with substrate. It is proportional to the concentration of enzyme, and measures the catalytic efficiency of the enzyme: the bigger the V_{max} the more efficient the enzyme.

Figure 2-3. Graphical Representation of the Michaelis-Menten Equation.

K_M, the **Michaelis constant**, or $(k_2 + k_3)/k_1$, is equal to the substrate concentration at which the reaction rate is half its maximal value (V_{max}), and is in units of moles/liter. A high K_M indicates weak binding between enzyme and substrate; when dissociation of ES complex to E and P is the rate-limiting step (ie., $k_1, k_2 >> k_3$), K_M becomes the dissociation constant of ES.

Lineweaver-Burk plot (Figure 2-4). When the reciprocal of velocity is plotted against the reciprocal of substrate concentration, the graph is in the form of $y = mx + b$, and the slope and y-intercept are easily obtained. They are K_M/V_{max} and $1/V_{max}$, respectively. The x-intercept is $-1/K_M$:

$$\frac{1}{v} = \frac{1}{V_{max}} + \frac{1}{[S]} \cdot \frac{K_M}{V_{max}}$$

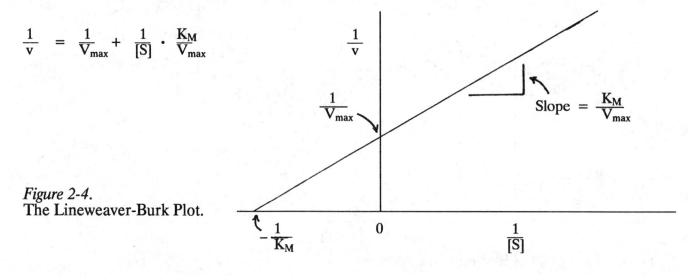

Figure 2-4.
The Lineweaver-Burk Plot.

C. Enzyme Inhibition

There are two major types: irreversible and reversible. Drugs are designed to inhibit specific enzymes in specific metabolic pathways.

1. *Irreversible inhibition* involves destruction or covalent modification of one or more functional groups of the enzyme. Examples:

   -Diisopropylfluorophosphate and other fluorophosphates bind irreversibly with the -OH of the serine residue of acetylcholine esterase.

   -Para-chloromercuribenzoate reacts with the -SH of cysteine.

   -Alkylating agents modify cysteine and other side chains.

   -Cyanide and sulfide bind to the iron atom of cytochrome oxidase.

   -Fluorouracil irreversibly inhibits thymidine synthetase.

   -Aspirin acetylates an amino group of the cyclooxygenase component of prostaglandin synthase.

2. *Reversible inhibition* (Figure 2-5) is characterized by a rapid equilibrium of the inhibitor and enzyme, and obeys Michaelis-Menten kinetics. There are three major types:

   a. **Competitive** inhibition: resembling the substrate, the inhibitor competes with it for binding to the active site of the enzyme.

Examples:	Inhibitor	Enzyme Inhibited
	malonate	succinate dehydrogenase
	sulfanilamide	dihydropteroate synthetase
	methotrexate	dihydrofolate reductase
	allopurinol	xanthine oxidase

   b. **Noncompetitive** inhibition: inhibitor does not resemble substrate, binds to the enzyme at a locus other than the substrate binding site. Examples: heavy metal ions — silver, mercury, lead, etc. Metalloenzymes are inhibited by metal-chelating agents that bind metal cofactors, e.g. EDTA.

   c. **Uncompetitive** inhibition: inhibitor binds to the enzyme-substrate complex and prevents further reaction.

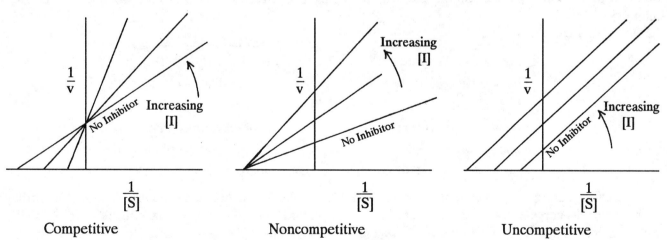

Competitive	Noncompetitive	Uncompetitive

Figure 2-5. Lineweaver-Burk Plots Showing the Effect of Inhibitor in Three Types of Enzyme Inhibition.

D. Active Site

The **active site** of an enzyme is the region that binds the substrate and contributes the amino acid residues that directly participate in the making and breaking of bonds. It is a three-dimensional entity having nonpolar clefts or crevices. These contain two types of amino acid residues: the **contact** (or catalytic) and the **auxiliary** amino acids. A substrate can induce a conformational change in the active site. This is known as the induced-fit model.

E. Catalytic Efficiency

A number of factors contribute to the efficiency of enzymes —

-Catalysis by distortion: a conformational change occurs in the substrate.

-A conformational change is induced in the enzyme.

-Covalent catalysis: a highly reactive covalent intermediate is formed.

-Acid and base catalysis.

-Proximity and orientation effect.

III. EFFECTS OF KINETIC PARAMETERS ON ENZYME ACTIVITY

A. Allosteric Enzymes

General properties. Enzyme activity is modulated through the noncovalent binding of a specific metabolite (allosteric effector, modulator or modifier) to the protein at a site **other than** the catalytic site (*allo* = other).

All known allosteric enzymes are oligomeric, ie., they have 2 or more polypeptide **subunits**, often four. Binding of the modulator to the allosteric site affects the binding of substrate to the catalytic site by changing the **quaternary structure** of the allosteric enzyme. The effect can be either **positive**, ie., to increase the binding of substrate, or **negative**, ie., to decrease the binding of substrate.

Regulation frequently occurs at the first, or **committed step** of a metabolic pathway, or at a **branch point**, with the final product of the pathway as a negative effector. This is end-product or feedback inhibition.

Examples:	Enzyme	Allosteric Effector
	homoserine dehydrogenase	threonine (–)
	homoserine succinylase	methionine (–)
	threonine deaminase	isoleucine (–)
	aspartate transcarbamoylase	cytidine triphosphate (–)
	phosphofructokinase	fructose-6-phosphate (+)
	pyruvate carboxylase	acetyl-CoA (+)

Kinetics (Figure 2-6). Allosteric enzymes do not follow Michaelis-Menten kinetics. A **sigmoidal** rather than hyperbolic curve is obtained when reaction rate is plotted against substrate concentration. This kinetic behavior is analogous to the binding of oxygen to hemoglobin. (Oxygen binding to myoglobin follows Michaelis-Menten kinetics instead.) The sigmoidal shape of the curve is caused by positive cooperativity, ie., binding of the first substrate molecule enhances the binding of subsequent molecules at other sites.

An activator has the effect of shifting the curve to the left; an inhibitor shifts it to the right. Thus the sigmoidal response of allosteric enzymes facilitates a more vigorous control of enzyme activity.

Mechanism of the Regulatory Activity. There are two major models proposed for cooperative binding of allosteric enzymes:

1. Symmetry-concerted model. Enzymes are generally composed of 2 or more identical subunits arranged in a symmetrical manner. If there is a change in conformation of one subunit, all other subunits must change their conformation in a concerted manner to preserve symmetry.

2. Sequential model. The binding of substrate changes the shape of the subunit to which it is bound. However, the conformation of the other subunits in the enzyme molecule are not appreciably altered. The conformational change caused by the binding of substrate in one subunit can increase or decrease the substrate binding affinity of the other subunits in the same enzyme molecule.

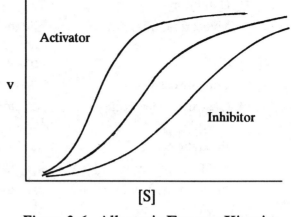

Figure 2-6. Allosteric Enzyme Kinetics.

B. Covalently Modified Regulatory Enzymes

Some regulatory enzymes may be modified by phosphorylation. Covalent binding of modifier (phosphate) to the enzyme changes its activity. Example: phosphorylation of active glycogen synthetase turns the enzyme into the inactive form, while phosphorylation of inactive glycogen phosphorylase turns the enzyme into the active form.

IV. GENERAL ASPECTS

A. Isozymes

Isozymes are enzymes that catalyze the **same reactions** and have similar molecular weights but **differ in subunit composition** and physical chemical properties. They may also differ in K_M, V_{max}, optimal temperature and pH, or substrate specificity. They often contain **multiple polypeptide subunits** of 2 or more types.

An example is *L-Lactate Dehydrogenase* (LDH). It is a tetramer of 2 types of subunit, M (muscle) and H (heart). Five isozymes occur: H_4, H_3M, H_2M_2, HM_3, and M_4. Various mixtures of isozymes are found in different tissues. H_4 occurs predominantly in cardiac tissue, and M_4 in skeletal muscle and liver.

B. Medical Aspects of Enzymology

As diagnostic tools. Enzymes or isozymes normally found intracellularly in various organs can be used as indicators of organ damage when they are found in blood. These enzymes, although always present in blood at low levels, are elevated far above normal in pathological conditions. Some examples are listed on the next page (Table 2-1).

As Laboratory Reagents. In **simple assays**, enzymes may be used for accurate determination of small quantities of blood constituents. In **coupled assays**, combinations of enzymes are often used to measure concentrations of specific substrates, coenzymes, or products. For example, an enzyme reaction may be coupled to the conversion of NADH to NAD^+, the removal or production of which can be followed easily by measuring the absorbance of the solution spectrophotometrically.

If the assay is for substrate (or product), then enzyme, coenzymes, etc. must be in excess; if the assay is for enzyme, then substrate, etc. must be in excess.

Condition	Enzymes with Elevated Levels in Blood
Myocardial Infarction	glutamic-oxaloacetic transaminase (SGOT) lactic dehydrogenase — H_4 and H_3M isozymes creatine phosphokinase
Bone disease	alkaline phosphatase
Obstructive liver disease	sorbitol dehydrogenase lactic dehydrogenase M_4 and M_3H isozymes
Prostatic cancer	acid phosphatase
Acute pancreatitis	amylase
Muscular dystrophy	aldolase glutamic pyruvate transaminase (SGPT)

Table 2-1. Some Enzymes Useful in Diagnosis.

V. REVIEW QUESTIONS ON ENZYMES

> ***DIRECTIONS:*** For each of the following multiple-choice questions (1 - 27), choose the ONE BEST answer.

The diagram below refers to ***Questions 1-3***:

The conversion of succinate to fumarate is catalyzed by succinate dehydrogenase. The following results were obtained when the reaction was studied in the presence or absence of the competitive inhibitor malonate:

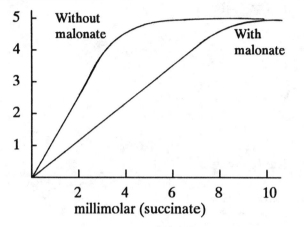

1. Without inhibitor the V_{max} is

A. 1 μmole/min
B. 2 μmoles/min
C. 3 μmoles/min
D. 4 μmoles/min
E. 5 μmoles/min

2. Without inhibitor the K_M for succinate is

A. 0.5 millimolar
B. 1 millimolar
C. 1.5 millimolar
D. 2 millimolar
E. 4 millimolar

3. The effect of malonate

A. is to increase ΔG of the reaction.
B. is to decrease the V_{max} to 2.5 μmole/min
C. is to increase the K_M to 6 millimolar
D. can be overcome by increasing the concentration of succinate.
E. results in cooperative binding of succinate.

The diagram below refers to ***Questions 4-9***:

The following graph contains several lines, each marked with a letter which represents a possible answer to a question. Line A represents the Lineweaver-Burk plot for the reaction of a normal substrate in the absence of any inhibitor.

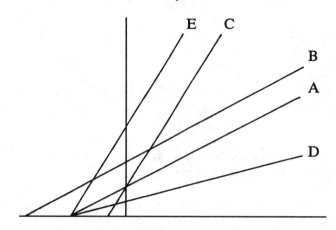

4. Which line would be expected in the presence of an uncompetitive inhibitor?

5. Which line would be expected in the presence of a noncompetitive inhibitor?

6. Which line would be expected in the presence of a competitive inhibitor?

7. Which line would be expected if the concentration of enzyme is doubled?

8. Which line has the same V_{max} as Line A but a larger K_M?

9. Which line has the same K_M as Line A but a smaller V_{max}?

10. Reactions catalyzed by an allosteric enzyme follow:

A. Michaelis-Menten kinetics
B. Lineweaver-Burk plot
C. Sigmoidal curve
D. Scatchard plot
E. Zymogen activation.

11. Lactic dehydrogenase is a tetrameric enzyme of 2 types of subunits, H and M, which associate according to the tissue of origin. The number of different tetrameric forms possible is

A. 1
B. 2
C. 3
D. 4
E. 5

12. Which of the following statements is NOT true for the active site of an enzyme?

A. It contains both contact and auxiliary amino acids.
B. It provides an environment favorable for the binding of the substrate and the enzyme.
C. It is maintained in proper conformation by the total 3-dimensional structure of the enzyme molecule.
D. It may contain, in addition to amino acid side chains, nonprotein constituents essential for catalytic action of the enzyme.
E. It must contain cysteine residues to form disulfide bonds with the substrate.

13. Data that would be useful in developing an assay for a serum enzyme of clinical interest include all of the following EXCEPT:

A. the molecular weight of the pure enzyme
B. K_M values for substrates so that they can be added to the assay in appropriate amounts
C. the occurrence of activation or inhibition by the substrate
D. the stability of the enzyme under various conditions
E. the optimal pH of the enzyme.

Questions 14-15:

The following data were obtained for an enzyme:

[Subs.] in Moles	Relative velocity (μmole prod/min)
1×10^{-4}	10
2×10^{-4}	20
3×10^{-4}	30
6×10^{-4}	55
9×10^{-4}	60
12×10^{-4}	60
15×10^{-4}	60

14. The V_{max} is

A. 10
B. 20
C. 30
D. 55
E. 60

15. The K_M is about

A. 1×10^{-4} M
B. 2×10^{-4} M
C. 3×10^{-4} M
D. 4.5×10^{-4} M
E. 6×10^{-4} M

16. The effect of pH on an enzyme-catalyzed reaction reflects:

A. increased rate of forward reaction
B. increased rate of reverse reaction
C. decreased activation energy for forward reaction
D. decreased activation energy for reverse reaction
E. ionization of the enzyme.

17. Conversion of zymogen to the active enzyme involves:

A. phosphorylation
B. methylation
C. dimer formation
D. cleavage of one or more specific peptide bonds
E. disulfide bond formation.

18. In catalyzing a reaction, an enzyme:

A. shifts the equilibrium toward product formation
B. lowers the energy of activation
C. shifts the equilibrium toward substrate formation
D. increases ground free energy of substrate
E. increases ground free energy of product.

19. Alpha-ketoglutarate dehydrogenase requires thiamin pyrophosphate as:

A. coenzyme
B. substrate
C. product
D. enzyme
E. holoenzyme.

20. In non-competitive inhibition, the

A. Inhibitor resembles the substrate
B. Substrate binds to the allosteric site
C. Inhibitor binds to the substrate-binding site
D. Inhibitor binds to a locus other than the substrate binding site
E. Inhibitor does not bind the enzyme at all.

21. Which of the following does NOT exist in precursor form?

A. carboxypeptidase
B. chymotrypsin
C. pepsin
D. insulin
E. transcarbamoylase.

22. The rate at which products are formed in a biosynthetic pathway is independent of the:

A. concentration of substrate
B. concentration of enzyme
C. temperature of the reaction
D. pH of the reaction environment
E. concentration of the zymogen.

23. The effect of a catalyst on a reaction is to:

A. increase the energy of activation
B. decrease the energy of activation
C. increase the ΔG of the reaction
D. decrease the ΔG of the reaction
E. increase the equilibrium constant.

24. K_M of an enzyme represents the

A. catalytic efficiency
B. substrate affinity
C. enzyme activity
D. cofactor affinity
E. coenzyme affinity.

25. V_{max} of an enzyme measures the

A. substrate affinity
B. catalytic efficiency
C. enzyme activity
D. cofactor affinity
E. coenzyme affinity.

26. Lactic dehydrogenase isozymes differ in

A. amino acid sequence
B. K_M for pyruvate
C. number of subunits
D. A and B are correct
E. A and C are correct.

27. One of the following is NOT an assumption in deriving the Michaelis-Menten equation:

A. An enzyme-substrate (ES) complex is an intermediate in the reaction.
B. There is a catalytic and a regulatory subunit in the enzyme.
C. The ES complex breaks down to reform the enzyme and substrate again.
D. The ES complex breaks down to form the product and the enzyme.
E. The formation of the ES complex is at equilibrium.

MATCHING: For each set of questions, choose the ONE BEST answer from the list of lettered options above it. An answer may be used one or more times, or not at all.

Questions 28-31: Blood levels of enzymes found intracellularly in various organs are used as indicators of organ damage. For each disease named in Questions 28-31, select the lettered enzyme with which it is most closely associated.

    A. Glutamic-oxaloacetic transaminase
    B. Glutamic-pyruvic transaminase
    C. Alkaline phosphatase
    D. Acid phosphatase
    E. Sorbitol dehydrogenase
    F. Ornithine transcarbamoylase
    G. Ornithine decarboxylase
    H. Glucose-6-phosphate dehydrogenase

28. Bone disease.

29. Muscular dystrophy.

30. Myocardial infarction.

31. Obstructive liver disease.

Questions 32-35:

    A. Biotin
    B. NAD
    C. FAD
    D. Pyridoxal phosphate
    E. GTP
    F. Coenzyme A
    G. ATP
    H. TPP

32. Coenzyme of pyruvate carboxylase.

33. Coenzyme of lactate dehydrogenase.

34. Coenzyme of glutamic-oxaloacetic transaminase

35. Coenzyme of amino acid oxidase.

Questions 36-44:

    A. Diisopropylfluorophosphate
    B. Parachloromercuribenzoate
    C. Sulfanilamide
    D. Fluorouracil
    E. Cyanide
    F. Malonate
    G. Aspirin
    H. Sulfanilamide
    I. Methotrexate
    J. Allopurinol

36. Inhibitor of thymidine synthetase.

37. Inhibitor of acetylcholine esterase.

38. Inhibitor of enzyme with catalytically essential sulfhydryl group.

39. Inhibitor of cytochrome oxidase.

40. Inhibitor of xanthine oxidase.

41. Inhibitor of dihyropteroate synthase

42. Inhibitor of succinate dehydrogenase

43. Inhibitor of dihydrofolate reductase

44. Inhibitor of cyclooxygenase.

Questions 45-50 are the numbered axes in the graphs below. For each question, select the item from the lettered list that provides the most appropriate label for the axis specified in the question. An item may be used one or more times, or not at all.

A. Reaction velocity
B. Concentration of substrate
C. Concentration of enzyme
D. Temperature
E. pH
F. Concentration of inhibitor
G. Concentration of coenzyme
H. Concentration of cofactor

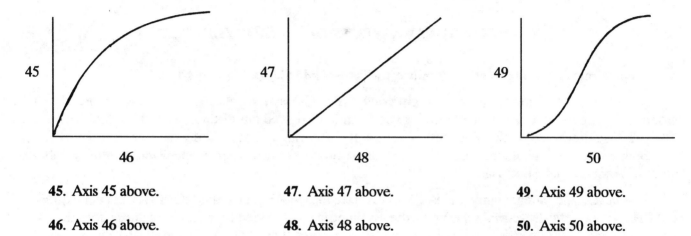

45. Axis 45 above.   **47.** Axis 47 above.   **49.** Axis 49 above.

46. Axis 46 above.   **48.** Axis 48 above.   **50.** Axis 50 above.

VI. ANSWERS TO QUESTIONS ON ENZYMES

1. E	11. E	21. E	31. E	41. H
2. D	12. E	22. E	32. A	42. F
3. D	13. A	23. B	33. B	43. I
4. B	14. E	24. B	34. D	44. G
5. E	15. C	25. B	35. C	45. A
6. C	16. E	26. D	36. D	46. B
7. D	17. D	27. B	37. A	47. A
8. C	18. B	28. C	38. B	48. C
9. E	19. A	29. B	39. E	49. A
10. C	20. D	30. A	40. J	50. B

3. CARBOHYDRATES

Robert E. Hurst and Albert M. Chandler

I. INTRODUCTION

Carbohydrate metabolism is the core of intermediary metabolism, providing a large part of the energy requirements for the organism, short-term storage of energy in the form of glycogen, and carbon skeletons for biosynthesis of other compounds. Amino acids and components of the citric (tricarboxylic) acid cycle feed into the pathways of carbohydrate metabolism. The following material is a concentrated essence of carbohydrate metabolism organized to show the pathways and their regulation.

II. STRUCTURAL ASPECTS OF CARBOHYDRATES

A. Carbohydrates fall into two major families, the ketoses and aldoses (Table 3-1).

   1. *Aldoses*. The parent compound is **glyceraldehyde**. C-2 is an asymmetric carbon, therefore, the OH group can be on the left or the right, forming two families of related but distinct D and L sugars. There is a total of 30 aldoses, 15 of the D-form and 15 of the L-form. Those aldoses most commonly seen in metabolism or in biological structures are **all of the D-form** and include **glyceraldehyde, erythrose, ribose, glucose, mannose** and **galactose**.

   2. *Ketoses*. the parent compound is **dihydroxyacetone** (DHA). Since the 2-C of DHA is not asymmetric, DHA is not optically active. Only when the asymmetric C-3 is added to make **erythrulose** does the compound become optically active. Thus there are fewer ketoses than there are aldoses. **D-Ribulose** and **D-xylulose** are found in the pentose phosphate pathway. **D-Fructose** is also an important hexose of this group.

	Aldoses	Ketoses
3C	Glyceraldehyde	Dihydroxyacetone
4C	Erythrose	Erythrulose
	Threose	
5C	Ribose	Ribulose
	Arabinose	Xylulose
	Xylose	
	Lyxose	
6C	Allose	Psicose
	Altrose	Fructose
	Glucose	Sorbose
	Mannose	Tagatose
	Gulose	
	Idose	
	Galactose	
	Talose	

Table 3-1. Aldoses and Ketoses through Six Carbons.

- 33 -

With the exception of DHA, all of these sugars are optically active, that is, in solution they turn the plane of polarized light either clockwise or counterclockwise (+ or –). There is **NOT** necessarily a correlation between the absolute configuration (D or L) and the direction of optical activity (+ or –).

B. Cyclization of Sugars (Figure 3-1)

An alcohol can react with an aldehyde to form a **hemiacetal**. When the aldehyde group of a hexose forms an internal hemiacetal with its C-5 hydroxyl, the cyclic compound formed is call a **pyranose**.

Ketones can also react with alcohols to form a **hemiketal**. In a pentose the C-2 ketone can form an internal hemiketal with the C-5 hydroxyl to form a **furanose**. In a hexose, it can also react with the C-6 hydroxyl to form a **pyranose**.

When sugars are depicted as their Haworth projections, they can be looked at "edge on". The formation of a hemiacetal or hemiketal creates a new asymmetric center (anomeric carbon) at C-1 or C-2. The OH group produced can be either above or below the ring. If it is projected **below** the ring it is in the α-position; **above** the ring, the β-position.

In solution, the linear and ring forms are in equilibrium. Which form predominates depends upon the sugar involved. Fructose prefers to form pyranose rings and this is its predominant form. Pentoses form only furanose rings.

The α and β forms of these sugars are also in equilibrium. Glucose, for example, exists in solution in an equilibrium mixture of about 30% α and 70% β. The open form is less than 1%. The form that predominates is sugar specific. There are enzymes that accelerate the changes between the α and β forms (**mutarotases**).

Conformation of the rings. The ring forms of the pyranoses and the furanoses are not planar. They form "boat" or "chair" forms in the case of pyranoses and "envelope" forms for furanoses. This is important for recognition purposes on the carbohydrate side chains of glycoproteins and glycolipids.

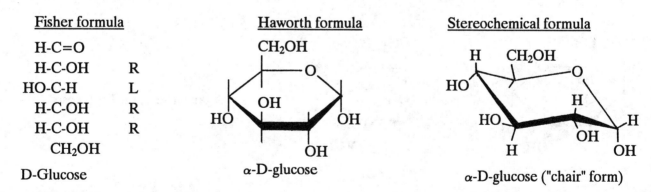

Figure 3-1. Conformation of Glucose, Depicted in Different Ways.

C. Sugar Derivatives

Glycosides are formed by splitting out a hydroxyl group between an alcohol and the -OH of a hemiacetal to form an **acetal**. This can be done either chemically or enzymatically. The bond formed is called a glycosidic bond and in this case is an O-glycosidic bond. N-glycosidic bonds can also be formed. The structures shown (Figure 3-2) are methylglycosides, α below the ring, β above.

This is a common way for sugars to link together. Cellobiose, for example, is a disaccharide composed of two glucoses linked (1,4) in a β linkage. It is a glycoside. If we string many glucoses together in this exact way we have cellulose.

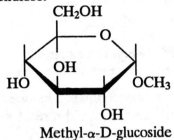

Methyl-α-D-glucoside

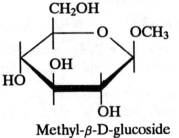

Methyl-β-D-glucoside

Figure 3-2. Sugar Glycosides.

An example of an **N-glycosidic** bond is adenosine. All nucleosides are made up of a base linked to either ribose or deoxyribose in β-N-glycosidic linkages. (See Chapter 11)

The most common disaccharides of interest are shown in Figure 3-3. Sucrose (table sugar) [glucose-fructose (α-1,2)]; lactose (milk sugar) [galactose-glucose (β-1,4)]; and maltose [2 glucose (α-1,4)].

Note that lactose can have both α and β anomers, depending upon the configuration of the glucose, while sucrose cannot have anomers because it lacks an anomeric carbon atom.

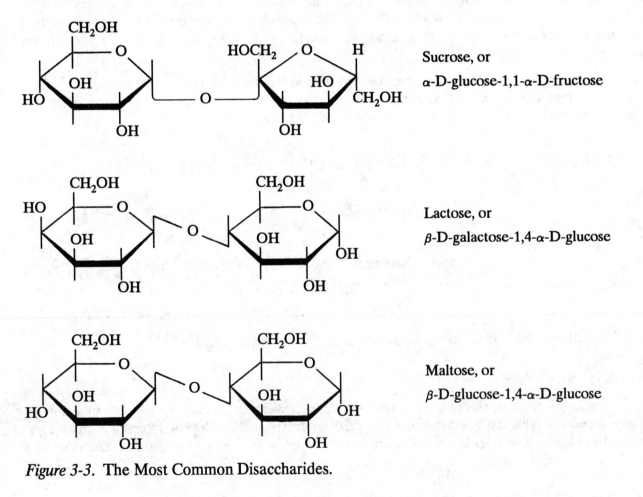

Sucrose, or

α-D-glucose-1,1-α-D-fructose

Lactose, or

β-D-galactose-1,4-α-D-glucose

Maltose, or

β-D-glucose-1,4-α-D-glucose

Figure 3-3. The Most Common Disaccharides.

III. DIGESTION

Carbohydrates are present in the diet as monosaccharides, disaccharides, and polysaccharides. The greatest amounts of glucose are ingested as starch (plant sources) and glycogen (animal sources).

Starch, the major dietary polysaccharide in the diet, exists in plants in two forms. Amylose, a long linear polymer of α-1,4 links and amylopectin, a structure consisting of regions of α-1,4 linked residues periodically branched by α-1,6 links (also called plant glycogen). Ingested starch is broken down by *α-amylases* secreted by the salivary glands and by the pancreas. This yields **maltose**, **maltotriose** and **oligosaccharides** which contain α-1,6 linkages. Other intestinal enzymes, *maltase* and *dextrinase*, complete the hydrolysis to glucose.

Disaccharides are cleaved in the small intestine by disaccharidase enzymes of varying specificities to monosaccharides. Examples include *maltase*, which cleaves maltose to **glucose**, *lactase*, which cleaves lactose to **glucose** and **galactose** and *sucrase* which cleaves sucrose to **glucose** and **fructose**. The resulting monosaccharides are then absorbed through the intestine into the bloodstream. There are several genetic diseases resulting from a deficiency in one of the disaccharidases, the most common one being lactase deficiency.

Lactase Deficiency. While infants of all races possess sufficient lactase in the intestine, adults of many races lack sufficient lactase, thereby giving rise to *lactose intolerance*. Intolerant individuals who consume more than a few ounces of milk will experience diarrhea and intestinal gas caused by microbial fermentation of lactose in the gut. Lactase persists in adults whose ancestors came from parts of the world with a long history of consuming milk, such as Northern Europe, parts of Central Asia, and parts of Africa. Most Blacks and Orientals are lactose-intolerant, as are significant proportions of Semitic and Mediterranean Caucasians.

Glycogen. Monosaccharides, especially glucose, are a readily available source of energy. Free glucose molecules cannot be stored efficiently in cells; therefore, they are condensed together as highly branched, insoluble structures; glycogen in animals, starch in plants. In glycogen the glucose molecules are linked together by α-1,4 and α-1,6 bonds. Glucose is freed from glycogen by the combined action of phosphorylase and debranching enzymes.

Cellulose. Cellulose is the most abundant polysaccharide in the world. *Animals do not synthesize cellulases (enzymes capable of hydrolyzing β-1,4 links) and, therefore, cannot utilize cellulose directly.* High fiber diets are mainly cellulose. Termites harbor protozoa which produce cellulases and easily digest cellulose. Animals with four stomachs (ruminants) support bacteria which are capable of digesting cellulose. The ruminants utilize the liberated glucose as a source of energy.

Glycoproteins. Glycoproteins are proteins which have oligosaccharide side chains covalently attached to the polypeptide backbone. (See the chapter on Protein Synthesis).

Glycolipids. Glycolipids are lipid molecules to which oligosaccharides are covalently linked. They are usually found in membranes with the carbohydrate moieties projecting out into the surrounding environment. Many cell-surface antigens or recognition sites are glycolipids. The oligosaccharide side chains are similar in structure to those found on glycoproteins.

Glycosaminoglycans. These compounds are complex polysaccharides found primarily in connective tissues where they form covalent links to proteins. The protein-carbohydrate compounds are called **proteoglycans** and are about 95% polysaccharide and 5% protein. Glycosaminoglycans are composed of multiple repeating units of disaccharides. One of the sugars is an amino sugar, the other is often a uronic acid. Some examples are shown in Table 3-2.

Several diseases exist that are the result of the deficiency of one or more enzymes responsible for the degradation of glycosaminoglycans.

Compound	Sugars
Chondroitin-6-sulfate	Glucuronic acid, GalNAc
Keratan sulfate	Galactose, GlcNAc
Heparin	Iduronic acid, Glucuronic acid GlcN-SO_4
Dermatan sulfate	Iduronic acid, GalNAc
Hyaluronic acid	Glucuronic acid, GlcNAc

Table 3-2. Some Glycosaminoglycans.

The role of these compounds is linked to their chemistry and structure. The repeated, closely spaced negative charges at physiological pH tend to cause the molecules to form stiff rods which form viscous solutions. They also bind divalent cations readily and tend to be highly hydrated and can act as lubricating agents.

IV. GLYCOLYSIS

In the text that follows, many abbreviations are used, summarized in Table 3-3 below.

Abbreviation	Name of Intermediate	Abbreviation	Name of Intermediate
1,3-BPG	1,3-Bisphosphoglyceric acid	G-3-P	Glyceraldehyde 3-phosphate
2,3-BPG	2,3-Bisphosphoglyceric acid	G-6-P	Glucose 6-phosphate
2-PG	2-Phosphoglyceric acid	Gal-1-P	Galactose 1-phosphate
3-PG	3-Phosphoglyceric acid	OAA	Oxaloacetate
DHAP	Dihydroxyacetone phosphate	PEP	Phosphoenolpyruvate
E-4-P	Erythrose 4-phosphate	R-5-P	Ribose 5-phosphate
F-1,6-bisP	Fructose 1,6-bisphosphate	S-7-P	Sedoheptulose 7-phosphate
F-1-P	Fructose 1-phosphate	UDPG	Uridyl diphosphoglucose
F-2,6-bisP	Fructose 2,6-bisphosphate	UDPGal	Uridyl diphosphogalactose
F-6-P	Fructose 6-phosphate	X-5-P	Xylulose 5-phosphate

Table 3-3. Abbreviations of Some Common Intermediates

Monosaccharides released by intestinal degradation of starches and other complex carbohydrates are rapidly absorbed in the small intestine. There are special transport systems for them. Glucose "carriers" in the plasma membranes of intestinal cells bind both glucose and sodium ion. Both are transported across the membrane; Na^+ down its concentration gradient, glucose against. ATP is utilized to pump Na^+ out again exchanging it for K^+. This is an example of **active transport**. The glucose and fructose carriers are rapidly saturated (see chapter on Membranes). The monosaccharides enter the bloodstream and are rapidly carried throughout the organism where they come into contact with all cells of the body. By far the most important sugar in terms of abundance is **glucose**, which may require insulin for uptake by some tissues (Table 3-4).

Readily permeable	Require Insulin
liver	muscle (all types)
brain	adipose tissue
lens	leukocytes
retina	mammary gland
erythrocytes	
kidney	

Table 3-4 Entry of Glucose into Cells

When the glucose molecule enters the cell it is acted upon by the **Glycolytic Pathway (Embden-Meyerhof Pathway)**. This pathway converts glucose to pyruvate (lactate) giving a net yield of 2 moles of ATP for every mole of glucose processed. The process is called **glycolysis**.

A. Overview of Glycolysis

1. All intermediates after glucose are phosphorylated and remain trapped in the cell. The end-product lactate is quite permeable and can go in and out.

2. Glucose is broken into two trioses which are interconvertible.

3. ATP is involved 4 times; NAD^+ once.

4. The key rate-limiting enzyme and the major point of control is *phosphofructokinase (PFK)*. Control points also exist at *hexokinase* and *pyruvate kinase*.

B. Reactions of Glycolysis (Figure 3-4)

1. *Hexokinase*, an enzyme with a high affinity for glucose (low K_M), catalyzes the phosphorylation of the hydroxyl at C-6. In liver and adipose tissue only, a second enzyme exists called *glucokinase*. It has a low affinity (high K_M) and comes into play when the blood levels of glucose are high and the capacity of hexokinase is saturated; that is, when conditions are favorable for storing excess energy as glycogen and fat. Glucokinase is inducible and the induction is insulin-dependent. The product of either kinase, **glucose-6-phosphate (G-6-P)**, can enter into several different pathways. (For abbreviations, see Table 3-3).

2. *Phosphoglucose Isomerase* converts G-6-P (an aldose) to **fructose-6-phosphate (F-6-P)** (a ketose). This is a freely reversible reaction.

3. *Phosphofructokinase (PFK)* catalyzes: F-6-P + ATP $\longrightarrow$ **F-1,6 bisphosphate**

This is a complex, multisubunit enzyme subject to control by allosteric regulation and phosphorylation. It is also inducible. In the older literature the product was referred to as a diphosphate but the more accurate term is bisphosphate.

4. *Aldolase* catalyzes a freely reversible reaction, splitting F-1,6-bisP into two 3-carbon units, **dihydroxyacetone phosphate (DHAP)** and **glyceraldehyde-3-phosphate (G-3-P)**. These trioses are closely related to one another, one being a ketose and the other an aldose.

5. The two trioses are readily converted one to the other by the enzyme *triose phosphate isomerase*. The strategy here is to convert all of the DHAP over to G-3-P which is the triose that is metabolized further.

6. *Glyceraldehyde 3-phosphate Dehydrogenase* catalyzes an oxidation-reduction step.

The first ATP generated by glycolysis ultimately comes from this reaction. G-3-P interacts with NAD^+ and inorganic phosphate to form **1,3-bisphosphoglyceric acid (1,3-BPG)** and $NADH + H^+$. Note that 1,3-BPG is a mixed anhydride and has a high energy bond with a $\Delta G^{\circ\prime}$ of hydrolysis greater than that for ATP. Arsenite can replace phosphate in this reaction but no ATP will be formed. The NADH can enter the electron transport system to form an additional 3 ATP under aerobic conditions.

7. *Phosphoglycerate Kinase* then catalyzes the reaction of 1,3-BPG with ADP to form **ATP** and **3-phosphoglyceric acid (3-PG)**.

8. *Phosphoglyceromutase* converts 3-PG to **2-PG**. This interconversion requires the presence of catalytic amounts of the intermediate, **2,3-BPG**. This latter compound also plays an important regulatory role in oxygen transport by hemoglobin. (See chapter on Amino Acids and Proteins).

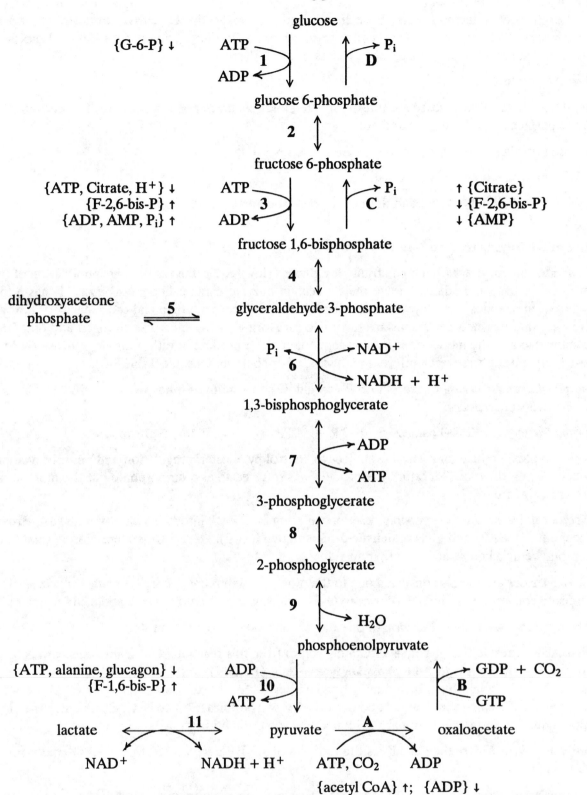

Figure 3-4. Glycolysis - Gluconeogenesis. Letters and numbers in boldface identify enzymes; see correspondingly numbered headings in text, next page. A - D are enzymes of gluco-neogenesis; see Section IX. Items in curved brackets refer to regulatory effects: ↑ = stimulatory; ↓ = inhibitory.

9. *Enolase* converts 2-PG into a second high-energy intermediate, **phosphoenolpyruvate (PEP)**. Dehydration instead of oxidation creates a high energy enol phosphate group. Resonance states have been increased in number and when the phosphate group is transferred, the enol converts to the favored ketone state driving the reaction to completion.

10. *Pyruvate Kinase* transfers a phosphate from PEP to ADP to form **pyruvate** plus a second **ATP**. This last reaction is essentially irreversible and drives the glycolytic pathway to completion. The formation of pyruvate is considered to be the end point of the glycolytic pathway.

V. ENTRY OF OTHER HEXOSES INTO THE GLYCOLYTIC PATHWAY

A. Fructose (Figure 3-5)

Most of the fructose in the diet comes from the hydrolysis of sucrose obtained from cane or beet sugar. Some fruits also contain considerable quantities of free fructose.

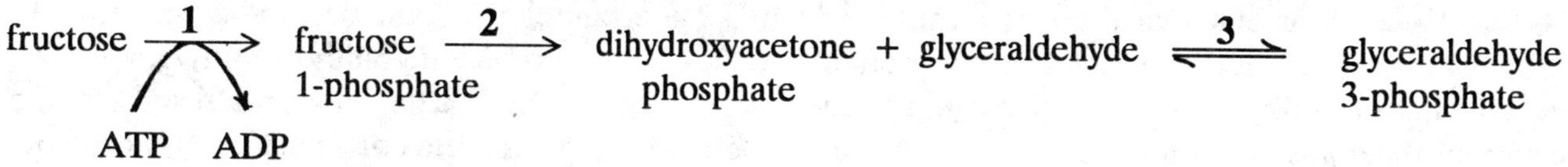

fructose $\xrightarrow{1}$ fructose 1-phosphate $\xrightarrow{2}$ dihydroxyacetone + glyceraldehyde $\underset{3}{\rightleftharpoons}$ glyceraldehyde 3-phosphate

phosphate

ATP ADP

Figure 3-5. Metabolism of Fructose. Numbers in boldface identify enzymes (see numbered sections below).

1. *Fructokinase.* Fructose can be phosphorylated by hexokinase but the K_M is extremely high so this is probably not a significant reaction. Fructokinase, found in liver, kidney and intestine uses ATP to form **fructose-1-phosphate (F-1-P)**. This is probably the major entry point for fructose.

Fructokinase will not phosphorylate glucose. It is not affected by insulin or fasting and fructose is metabolized at a normal rate by diabetics. Because fructose bypasses PFK it is metabolized more rapidly in the liver than is glucose. A genetic deficiency of fructokinase leads to **essential fructosuria**, a benign disorder.

2. *Fructose-1-phosphate Aldolase* (aldolase B, aldolase 2) cleaves F-1-P into DHAP and **glyceraldehyde**.

A deficiency of aldolase B results in **hereditary fructose intolerance**, a disease characterized by fructose-induced hypoglycemia and liver damage. In the presence of fructose, the resulting high level of fructose 1-phosphate inhibits liver phosphorylase, stopping glucose production from glycogen (see glycogen metabolism). It also sequesters all of the cell's phosphate and virtually stops ATP synthesis in the liver, which apparently prevents the maintenance of normal ionic gradients and leads to osmotic damage to hepatocytes.

3. *Triose Kinase* catalyzes the phosphorylation of glyceraldehyde by ATP. The G-3-P formed can either be converted to pyruvate or to glucose.

B. Galactose: obtained primarily by the ingestion of lactose in milk. (Figure 3-6)

1. *Galactokinase* converts galactose to **galactose-1-phosphate (Gal-1-P)**. ATP is the phosphate donor.

2. *Galactose-1-phosphate Uridyltransferase* catalyzes the reaction between Gal-1-P and **uridyl diphosphoglucose (UDPG)** to form **UDPGal** and **G-1-P**. This is a reversible exchange reaction: the UDP group is transferred from glucose to galactose, and the glucose 1-phosphate can be converted (*phosphoglucomutase*) to the glycolytic intermediate, glucose 6-phosphate.

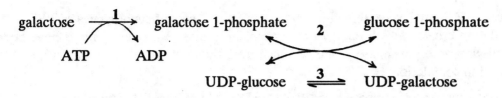

Figure 3-6. Metabolism of Galactose. Boldface numbers identify enzymes; (see numbered sections in text).

3. *UDPGal-4-epimerase* catalyzes the epimerization of UDPGal to UDPG, regenerating the UDP-glucose consumed by reaction (2), making a cycle. The reaction is reversible and requires NAD^+ as cofactor.

The UDPG formed from UDPGal can also enter the glycogen cycle and the glucose can be transferred to the non-reducing end of a glycogen molecule by glycogen synthase. When the glycogen is degraded the glucose is then converted to G-1-P by phosphorylase and then to G-6-P by phosphoglucomutase and then on to pyruvate by the glycolytic pathway.

Galactosemia is an autosomal recessive trait in which G-1-P accumulates in the cell and is converted to galactitol which collects in the lens and causes cataracts. Other symptoms include failure to thrive, vomiting and diarrhea after drinking milk, liver damage, and mental retardation. Severe **galactosemia** results from a deficiency of the *uridyltransferase* which allows the accumulation of large amounts of galactose 1-phosphate. It can also be caused,in a relatively mild form, by a deficiency of *galactokinase*.

C. Glycerol

Released from adipose tissue and liver primarily as the result of the hydrolysis of triglycerides.

1. *Glycerol kinase* catalyzes the following reaction, but its activity is low in muscle and adipose tissue:

$$\text{glycerol} + \text{ATP} \longrightarrow \text{glycerol-3-P} + \text{ADP}$$

2. *Glycerol phosphate dehydrogenase* catalyzes the reaction:

$$\text{glycerol-3-P} + NAD^+ \longrightarrow \text{dihydroxyacetone phosphate} + NADH + H^+$$

The dihydroxyacetone phosphate can then enter the glycolytic pathway to be converted to pyruvate or can be converted to glucose via the gluconeogenic pathway.

VI.   METABOLISM OF PYRUVATE

The pyruvate formed by glycolysis can go in three major directions.

A. Alcohol

In yeast and other microorganisms grown under anaerobic conditions the pyruvate is converted to alcohol. *Pyruvate decarboxylase* converts the pyruvate to **acetaldehyde** and CO_2. Then *alcohol dehydrogenase* reduces the acetaldehyde to **ethanol** using NADH as a cofactor. The NAD^+ formed from this reaction can then go back to the G-3-P dehydrogenase step to keep glycolysis going.

B. Lactate

Erythrocytes lack mitochondria, therefore their metabolism is purely anaerobic. Under prolonged, intense activity muscles are unable to obtain sufficient oxygen to convert all of the pyruvate formed to CO_2 and water via the tricarboxylic acid cycle and shift over to anaerobic metabolism. Under anaerobic conditions both muscle and erythrocytes convert the pyruvate to **lactate**.

Formation of lactate from pyruvate is catalyzed by *lactate dehydrogenase* which uses NADH as a cofactor. The function of this conversion is to regenerate NAD^+ which can then go back to oxidize G-3-P to form two more ATP molecules.

C. Acetyl CoA

Under aerobic conditions pyruvate is converted to **acetyl CoA** by *pyruvate dehydrogenase*. The acetyl CoA can then enter the Tricarboxylic Acid Cycle. (See Chapter 4).

VII. REGULATION OF GLYCOLYSIS

A. Phosphofructokinase

Allosterically inhibited by high concentrations of ATP and of citrate. Hydrogen ion is also inhibitory. When the enzyme is saturated, F-6-P accumulates; excess F-6-P is shunted though a second pathway to form **fructose-2,6-bisphosphate (F-2,6-bisP)**. F-2,6-bisP is a potent *allosteric activator* of PFK. The regulation of synthesis of this compound is itself very complex but is primarily governed by the levels of F-6-P and ATP.

B. Hexokinase: High levels of G-6-P inhibit hexokinase allosterically.

C. Pyruvate kinase exists in three different forms (isozymes) in different tissues:

1. *Liver* (L-form): regulated by F-1,6-bisP (+) and ATP (–) and alanine (–). This form can also be phosphorylated (inactivated) and dephosphorylated (activated). Phosphorylation (inhibition) occurs when the cell is in the gluconeogenic mode.

2. *Muscle* (M-form): this enzyme is not phosphorylated because muscle does not carry out gluconeogenesis.

3. *Other tissues* (A-form): regulation is in between liver and muscle forms.

D. 2,3-Bisphosphoglyceric acid (2,3-BPG)

In red blood cells the glycolytic step from 1,3-BPG to 3-PG can be bypassed. An additional enzyme, *bisphosphoglycerate mutase*, converts 1,3-BPG to **2,3-BPG**. 2,3-BPG can, in turn, be converted to 3-PG by *2,3-BPG phosphatase*. One ATP is lost when glycolysis goes by this route. The energy normally trapped as ATP is dissipated as heat. The 2,3-BPG can then combine with hemoglobin to assist in oxygenating the peripheral tissues. (See chapter on Amino Acids and Proteins).

VIII. PENTOSE PHOSPHATE PATHWAY

A second major route for G-6-P is through the **Pentose Phosphate Pathway** (Pentose Shunt, Hexose Monophosphate Shunt). The major functions for this pathway are to produce **NADPH** for reductive biosyntheses and **pentoses** used primarily for nucleic acid synthesis. The pathway also provides a mixture of 3, 4, 5, 6, and 7-C sugars which may be used as precursors for other compounds. (These are often abbreviated; see Table 3-4).

A. Oxidative Phase (G-6-P to Ribulose-5-P). Three enzymes are involved:

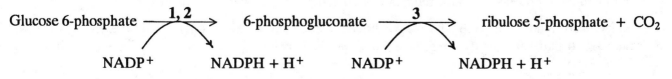

1. *G-6-P dehydrogenase* catalyzes the irreversible dehydrogenation of G-6-P to **6-P-gluconolactone**, forming an internal ester between C-1 and C-5. *This is the committed, rate-limiting step in the pathway.* The enzyme is highly specific for **NADP$^+$** and the level of this compound controls the rate of the reaction. High concentrations of NADPH will compete with NADP$^+$ binding and slow the reaction down.

2. *Lactonase* hydrolyzes 6-P-Gluconolactone to **6-P-gluconic acid**.

3. *6-P-gluconic acid dehydrogenase* also uses NADP$^+$ as an electron acceptor and converts 6-P-gluconate to **ribulose-5-P** and **CO_2**. The release of CO_2 drives the reaction to completion. For every G-6-P committed to this pathway, 2 NADPH, 1 CO_2 and 1 ribulose-5-P are obtained.

B. Non-oxidative Phase

Under most conditions the amount of pentoses formed is in excess of requirements; therefore, this excess is put back into glycolytic intermediates. This is accomplished by the non-oxidative phase of the pentose phosphate pathway, which converts 3 pentoses to 2 hexoses and 1 triose.

The non-oxidative branch consists of five reactions catalyzed by four enzymes and can be summarized as follows:

$$3 \text{ Ribulose 5-P} \underset{\quad}{\overset{1}{\rightleftharpoons}} 3 \text{ ribose 5-P} \underset{\quad}{\overset{2, 3, 4, 3a}{\rightleftharpoons}} 2 \text{ F-6-P} + 1 \text{ glyceraldehyde 3-P}$$

1. *Phosphopentose isomerase* converts ribulose-5-P to **ribose-5-P (R-5-P)**. R-5-P is a key intermediate for the formation of nucleosides, nucleotides and nucleic acids.

2. *Phosphopentose epimerase* can convert ribulose-5-P to **xylulose-5-P**.

3. *Transketolase* transfers the top two carbons from **xylulose-5-P (X-5-P)** to R-5-P to form G-3-P and **sedoheptulose-7-P (S-7-P)**. The X-5-P comes from the epimerization of ribulose-5-P. This enzyme requires thiamin pyrophosphate as cofactor.

4. *Transaldolase* transfers the top three carbons from S-7-P to G-3-P to form **erythrose-4-P (E-4-P)** + F-6-P.

3a. *Transketolase* also transfers the top two carbons from X-5-P to E-4-P to form G-3-P again and another F-6-P.

NOTE: Since these enzymes catalyze freely reversible reactions it is possible to synthesize R-5-P without going through the oxidative phase of the pathway.

C. G-6-P Dehydrogenase Deficiency and Hemolytic Anemia

Red blood cells contain high concentrations of glutathione. In its reduced form (free -SH) glutathione acts as a protective scavenging agent against hydrogen peroxide and other strong oxidizing agents. If these agents are not neutralized they will oxidize hemoglobin iron to the ferric state and oxidize membrane lipids. The hemoglobin will not carry oxygen and the membranes will become very fragile. Thus reduced glutathione helps to keep hemoglobin in the Fe^{++} state and helps to maintain red blood cell membrane integrity. It can be used up however, and when fully converted to G-S-S-G it no longer functions as a protective agent.

Red blood cells do not have mitochondria, thus their metabolism is mainly glycolytic. They do, however, have a very active pentose phosphate pathway. The NADPH generated by this pathway can be used to keep glutathione in the reduced state. This reduction is carried out by *glutathione reductase*.

A significant number of people have a sex-linked genetic defect that results in a deficiency in G-6-P dehydrogenase. It particularly affects blacks and like the sickle cell trait, apparently persists because it affords protection against the malarial organism. The erythrocytes of people with this deficiency have difficulty generating enough NADPH to keep the glutathione reduced. The red blood cells of these individuals are readily hemolyzed, particularly after exposure to certain drugs. These include antimalarials, aspirin and sulfonamides. Fava beans also contain a compound which is highly toxic to these cells. The hemolysis can be so great as to be fatal.

D. Adipose Tissue and Phagocytic Cells

The activity of the pentose phosphate pathway is very high in adipose tissue. A major purpose of this tissue is to synthesize fatty acids and store them as triglycerides. This synthesis requires the tissue to have a very active source of reducing power. For every molecule of fatty acid made from acetyl CoA about 14 - 16 NADPH are needed, about half of which comes from the pentose phosphate pathway. This pathway is also very active in cells undergoing **phagocytosis**.

IX. GLUCONEOGENESIS

Gluconeogenesis is the synthesis of glucose from non-carbohydrate precursors. These precursors are primarily lactate, glycerol and certain amino acids.

A. Function and Importance

This is an extremely important pathway for the maintenance and functioning of the **central nervous system**. The brain uses glucose as its primary fuel. Calculations have shown that the total daily body requirement for glucose is over 150 g/24 hours and the brain uses about 75% of this. The total body reserve in the form of glucose and glycogen is about 200 g or a little over one day's requirement. If one fasts for longer than a day or uses up these reserves faster through intense exercise, glucose must be supplied by the conversion of other compounds to glucose. (In prolonged fasting the supply of glucose is partially spared when the brain adapts to using ketone bodies).

The circulating blood glucose must be kept within fairly narrow limits. If the concentration drops too low (hypoglycemia) brain dysfunction occurs. If severe enough it can lead to coma and death. Glucose is also required for adipose tissue to produce glycerol for triglyceride formation and for the metabolism of erythrocytes. The concentrations of citric acid cycle intermediates can be maintained only if some glucose is being metabolized (amphibolic pathway). Thus there is a basal requirement for glucose even when most of the calories are supplied from triglycerides. Skeletal muscle operating under anaerobic conditions primarily uses glucose as a fuel and during intense exercise releases large amounts of lactate. Erythrocytes also release considerable quantities of lactate. Adipose tissue releases glycerol continuously. The lactate and glycerol must be recycled to glucose. Finally, under severe stress or starvation, glucose levels are maintained by breaking down skeletal muscle proteins so that some of the amino acids released can be converted to glucose. This recycling and conversion is carried out by the gluconeogenic pathway.

B. Energy Barriers to the Reversal of Glycolysis

Pyruvate is a common key compound for both glycolysis and gluconeogenesis. However, gluconeogenesis is **NOT** the reversal of glycolysis. In the glycolytic pathway, the conversion of PEP to pyruvate greatly favors the formation of pyruvate. There is a very large energy barrier inhibiting the reversal of this reaction. The formation of G-6-P and its conversion to F-1,6-bisP are also irreversible. Specific gluconeogenic enzymes make "end-runs" around these energy barriers.

1a. Pyruvate carboxylase [reaction (a), below, and enzyme A, Figure 3-4]

(a)	Pyruvate + CO_2 + ATP	$\longrightarrow$	OAA + ADP + P_i + 2 H^+
(b)	OAA + GTP	$\rightleftharpoons$	PEP + GDP + CO_2

Sum: Pyruvate + ATP + GTP $\longrightarrow$ PEP + ADP + GDP + P_i + 2 H^+

All carboxylations require **biotin** as a cofactor and pyruvate carboxylaseis a biotin-containing enzyme, binding CO_2 (HCO_3^-) in an "active" form. This activation step requires the utilization of 1 ATP. Pyruvate carboxylase transfers the activated CO_2 to pyruvate to form **oxaloacetate (OAA)**.

Acetyl CoA is an **obligatory allosteric effector** of pyruvate carboxylase. A high level of acetyl CoA is a signal that more OAA must be synthesized. Acetyl CoA links both glucose metabolism and fatty acid metabolism to the TCA cycle. If the concentrations of both acetyl CoA and ATP are high, the OAA will be directed toward glucose formation (gluconeogenesis). If acetyl CoA concentration is high but ATP is low, the OAA will be shunted to the TCA cycle to form CO_2 and water (ATP formation).

1b. Pyruvate Carboxykinase [reaction (b) and enzyme B, Figure 3-4]

The OAA is concurrently decarboxylated and phosphorylated by *PEP carboxykinase*. GTP donates the phosphate group. The removal of CO_2 drives the reaction to completion. Thus the sum of reactions (a) and (b) is the conversion of pyruvate to the high-energy compound, PEP.

2. Fructose-1,6,bisphosphatase [enzyme C, Figure 3-4] catalyzes the second major step toward glucose. The phosphate on the 1 position of F-1,6-bisP is cleaved forming F-6-P.

3. Glucose-6-phosphatase [enzyme D, Figure 3-4] catalyzes the hydrolysis of the phosphate to form free glucose. Glucose-6-phosphatase is membrane-bound ensuring that the glucose released leaves the cell.

NOTE: Glucose-6-phosphatase does NOT exist in brain, adipose tissue or muscle, therefore, these tissues are not gluconeogenic. The major gluconeogenic tissues are **liver**, **kidney** *and* **intestinal epithelium**.

C. Compartmentalization of the Reactions

All enzymes of glycolysis are cytosolic. All enzymes of the TCA cycle and of oxidative phosphorylation reside in mitochondria. Enzymes of the gluconeogenic pathway are in both: pyruvate carboxylase is mitochondrial and PEP carboxykinase is cytosolic. But the mitochondrial membrane is not permeable to OAA, an intermediate of the TCA cycle. However, it is permeable to **malate**, a derivative of OAA, and to pyruvate. Thus pyruvate in the cytosol enters mitochondria where it encounters pyruvate carboxylase and is converted to OAA. The OAA can be reduced to malate by *malate dehydrogenase* (NADH involved). The malate leaves the mitochondrion by means of an antiport system in exchange for one of several substrates (P_i, citrate, α-KG or other dicarboxylic acids) and is reoxidized back to OAA by a cytosolic form of malate dehydrogenase (NAD^+ involved). A second transport mechanism involves the transamination of oxaloacetate to aspartate which is exchanged by another antiport system for glutamate. The aspartate is transaminated back to OAA in the cytosol. The OAA is then acted upon by PEP carboxykinase to form PEP.

D. Energy Requirements and Regulation of Gluconeogenesis

Gluconeogenesis is an energy-expensive process. It takes six high energy bonds (4 ATP + 2 GTP) to make one molecule of glucose from 2 pyruvate molecules.

The pathways of glycolysis and gluconeogenesis are regulated reciprocally. It would make no sense to catabolize and synthesize glucose simultaneously, that is, run a "futile cycle". The key enzymes of glycolysis and gluconeogenesis, and their allosteric regulators are shown in Table 3-5.

Glycolysis	Gluconeogenesis
PFK	*F-1,6-bisPase*
ATP(−)	ATP(+)
AMP(+)	AMP(−)
citrate(−)	citrate(+)
F-2,6-bisP(+)	F-2,6-bisP(−)
Pyruvate kinase	*Pyruvate carboxylase*
F-1,6-bisP(+)	Acetyl CoA(+)
ATP(−)	ADP(−)

Table 3-5. Regulators of Glycolysis and Gluconeogenesis

E. Lactate Dehydrogenase and the Cori Cycle

Lactate dehydrogenase (LDH) catalyzes the interconversion of pyruvate to lactate and *vice versa*. LDH exists in several isozyme forms; each tissue has a distinct isozyme distribution. Analysis of which of the isozyme forms predominates in serum can indicate which tissue or organ has suffered significant damage.

When peripheral tissues undergo anaerobic metabolism they release lactate into the bloodstream. This is carried to the liver where it is converted back to glucose by gluconeogenesis. The glucose is released into the bloodstream and is again carried to the peripheral tissues. This cycle is referred to as the **Cori Cycle**.

F. The Alanine Cycle

Under conditions of stress, glucocorticoids cause the induction of gluconeogenic enzymes in the liver. They also cause the breakdown of muscle proteins, releasing amino acids into the bloodstream. These are carried to the liver where the glycogenic amino acids (characterized by alanine) are converted to glucose via the gluconeogenic pathway. The glucose is released into the bloodstream where it can be utilized by the peripheral tissues. This is known as the **Alanine Cycle**.

X. GLYCOGEN METABOLISM

A. Glycogen: Structure and Location

Glycogen consists of relatively long chains of glucose residues linked together by α-1,4 bonds. These chains are branched at about every 10 residues via α-1,6 bonds.

Sugars are referred to as reducing or non-reducing sugars. This arises from the fact that aldehydes or ketones in alkaline solutions are capable of reducing Cu^{++} to CuO, forming a reddish precipitate. If the C-1 carbon of an aldose is tied up in a glycosidic link it cannot act as a reductant. With regard to glycogen chains, therefore, one can refer to their reducing or non-reducing ends.

Glycogen is stored in cells, mainly liver and muscle, as granules. The enzymes responsible for glycogen metabolism are also bound to these granules. Several inherited deficiencies involving these enzymes are responsible for the several different glycogen storage diseases.

B. Glycogenolysis (Figure 3-7)

1. *Glycogen Phosphorylase*, by the process of **phosphorolysis**, catalyzes removal of glucose molecules one by one from the outermost ends of the chains (non-reducing ends), releasing each glucose residue as a molecule of G-1-P. Theoretically, this is a reversible reaction but in the cell the ratio of P_i / G-1-P is high

enough to drive the reaction only toward glycogen breakdown. **Pyridoxal phosphate (Vit B$_6$)** is a cofactor for this enzyme.

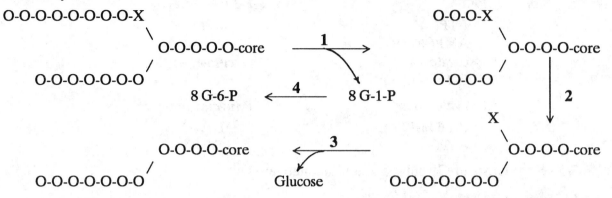

Figure 3-7. Glycogenolysis. Numbered steps correspond to numbered sections in text.

2. *Glucan Transferase*. Phosphorolysis of the glucose residues continues down the chain until the enzyme gets within 4 residues of an α-1,6 link. At this point a second enzyme, *glucan transferase*, removes three of the four residues, transferring them to the non-reducing end of another chain.

3. *Debranching enzyme (α-1,6 glucosidase)* removes the remaining residue. Phosphorylase can then continue down the chain until another branch point is reached. The transferase and debranching activities appear to be on the same polypeptide in separate domains.

4. *Phosphoglucomutase* (PGM) converts the released G-1-P to G-6-P. The G-1-P to G-6-P conversion requires the participation of an intermediate compound, **G-1,6-bisP**. The residue released by the debranching enzyme is free glucose, not G-1-P, and must be phosphorylated by hexokinase before being metabolized further.

The energy conservation from the breakdown of glycogen is 90 % efficient, since only about 1 residue in 10 is released in the form of free glucose which must be re-phosphorylated before it can be metabolized. Since the metabolism of 1 mole of G-1-P yields 37 moles of ATP but requires the expenditure of only 1 mole of ATP, the recovery of ATP-energy is over 97 %.

C. Glycogen Synthesis

$$\text{G-1-P} + \text{UTP} \xrightarrow{\;\;1\;\;} \text{UDPG} + \text{PP}_i$$

$$\text{Glycogen}_{(n)} + \text{UDPG} \xrightarrow{\;\;2\;\;} \text{glycogen}_{(n+1)} + \text{UDP}$$

1. *UDPG pyrophosphorylase* [(1) above] catalyzes production of UDPG, the activated form of glucose. The release of PP$_i$, which is readily hydrolyzed in the cell, drives this reaction toward UDPG formation.

2. *Glycogen Synthase* [(2) above] carries out the transfer of a glucose residue to the non-reducing end of a glycogen chain. A chain of at least four glucose residues, bound to protein, is required as a primer.

3. *Branching Enzyme* creates α-1,6 linkages. Fragments of chain of about seven residues long are broken off at an α-1,4 link and re-attached elsewhere through an α-1,6 link. The branching process increases the solubility of glycogen and also increases the number of terminal residues which can be attacked by phosphorylase and glycogen synthase.

D. Regulation of Glycogen Metabolism

Hormones play a vital role in the control of glycogen metabolism. Insulin stimulates glycogen synthesis in liver and muscle. Glycogenolysis is stimulated by epinephrine (mainly muscle) and glucagon (mainly liver).

Phosphorylation-Dephosphorylation. Both epinephrine and glucagon bind to cell membrane receptors that are linked to *adenyl cyclase*. This promotes the formation of **cyclic AMP** which in turn allosterically activates one or more protein kinases. These kinases then catalyze the phosphorylation of one or more target proteins using ATP as the phosphate donor. In the case of glycogen degradation, the target protein is *phosphorylase kinase*. Phosphorylation of this protein converts it from an inactive form to an active form. Activated phosphorylase kinase in turn phosphorylates inactive *phosphorylase b* converting it to *phosphorylase a*. This is the **active** form of phosphorylase.

The cyclic AMP-dependent protein kinase that activates phosphorylase kinase is also known as *phosphorylase kinase kinase*. This enzyme has broad specificity and can also phosphorylate *glycogen synthase*. In the case of glycogen synthase, the non-phosphorylated form is active (a) and the phosphorylated form is inactive (b). Thus this one enzyme can reciprocally control both the synthesis and breakdown of glycogen.

Allosteric Regulators (Muscle). Increased concentration of AMP can stimulate phosphorylase b allosterically. The initial contraction of muscle breaks down ATP to ADP. *Myokinase* takes 2 ADP and forms 1 ATP and 1 AMP. If the formation of ADP occurs rapidly enough, AMP concentrations will rise and will activate the phosphorylase b. On the other hand, high concentrations of ATP and G-6-P inhibit phosphorylase b activity by allosteric means.

Phosphorylase kinase is a multi-subunit enzyme. One of these subunits is **calmodulin**. Calmodulin has a great affinity for Ca^{++} and participates in the regulation of many different enzymes. Ca^{++} is released when muscle contracts and the increased Ca^{++} concentration binds to the calmodulin subunit partially activating phosphorylase kinase b allosterically.

Phosphatases. All of the phosphorylations described above can be reversed by specific *phosphatases*. *Protein phosphatase 1* converts phosphorylase a back to phosphorylase b. It also converts glycogen synthase b back to a.

Phosphatase 1 is blocked by a protein called **inhibitor 1**. Inhibitor 1 binds when it is phosphorylated but does not bind when dephosphorylated. The phosphorylation of inhibitor 1 is catalyzed by a specific cAMP-dependent protein kinase. Insulin decreases the amount of phosphorylated inhibitor 1.

Regulation of Liver Glycogen. Liver cells monitor the concentration of blood glucose. Liver phosphorylase a acts as a **blood glucose sensor**. When glucose binds to phosphorylase a it causes an allosteric change that exposes the usually cryptic or hidden phosphate group. *Phosphatase 1*, which is tightly bound to the phosphorylase a, then cleaves the phosphate off, inactivating the phosphorylase a by converting it to the b-form. This conversion releases the phosphatase 1, allowing it to now attack the inactive, phosphorylated glycogen synthase and converting it to its active form.

XI GLYCOGEN STORAGE DISEASES

A number of diseases of glycogen metabolism have been identified. Depending upon which enzyme is deficient, these diseases can manifest either storage of excessive levels of glycogen or the synthesis of glycogen of abnormal structure, or both. Their characteristics are summarized in Table 3-6.

Type	Name	Defective enzyme	Organ affected	Glycogen levels and structure *
I	von Gierke's Disease	Glucose 6-Pase	Liver & Kidney	I, N
II	Pompe's Disease	α-1,4-glucosidase	All organs	very I, N
III	Cori's Disease	debranching enzyme	Muscle and Liver	I, A: short outer branches
IV	Andersen's	branching enzyme	Liver and Spleen	I, A: v. long outer branches
V	McArdle'sDisease	phosphorylase	Muscle	I, N
VI	Hers'Disease	phosphorylase	Liver	I, N
VII		phosphofructokinase	Muscle	I, N
VIII		phosphorylase kinase	Liver	I, N

Table 3-6. Glycogen Storage Diseases: Nomenclature and Effects on Glycogen
*I = increased; N = normal levels or structure; A = abnormal structure.

XII. REVIEW QUESTIONS ON CARBOHYDRATES

> *DIRECTIONS:* For each of the following multiple-choice questions (1 - 21), choose the ONE BEST answer.

1. The immediate products of oxidation of one mole of glucose 6-phosphate through the oxidative portion of the pentose phosphate pathway are:

A. 2 moles of reduced NAD, one mole of ribulose 5-phosphate and one mole of CO_2
B. 2 moles of oxidized NADP, one mole of ribulose 5-phosphate and one mole of CO_2
C. 2 moles of reduced NADP, one mole of xylulose 5-phosphate and one mole of CO_2
D. 2 moles of reduced NADP, one mole of ribulose 5-phosphate and one mole of CO_2
E. one mole of fructose 6-phosphate and five moles of CO_2

2. A patient presenting with a suspected metabolic disorder shows (1) abnormally high amounts of glycogen with normal structure in liver, and (2) no increase in blood glucose levels following oral administration of fructose. From these two findings, which one of the following enzymes is likely to be deficient?

A. phosphoglucomutase
B. UDP-glycogen transglucosylase
C. fructokinase
D. glucose-6-phosphatase
E. glucokinase

3. The Cori Cycle, glucose → 2 lactate + 2 ATP (muscle) and 2 lactate + 6 ATP → glucose (liver), is important because:

A. there is a net destruction of ATP, restoring the energy balance between muscle and liver.
B. it results in the net generation of glucose in the liver and ATP in the muscles without the build up of high lactate levels.
C. it enables muscle mass to be used for energy in conditions of extreme starvation.
D. it serves to prevent lactate levels from dropping too low in the blood, which would impair brain function.
E. it enables G-6-P to be transported across the liver cell plasma membrane.

4. Which one of the following structures is D-glucose?

A.	B.	C.	D.	E.
HC=O	HC=O	HC=O	CHO	CHO
HOCH	HCOH	HCOH	C=O	C=O
HOCH	HOCH	HOCH	HOCH	HCOH
HCOH	HCOH	HOCH	HCOH	HOCH
HCOH	HCOH	HCOH	HCOH	HCOH
CH_2OH	CH_2OH	CH_2OH	CH_2OH	CH_2OH

5. Glycolysis in muscle is reduced when fatty acid oxidation is increased because:

A. oxidation of fatty acids increases the formation of citrate which blocks phosphofructokinase.
B. utilization of fatty acid increases the level of NADPH which decreases the activity of hexokinase.
C. fatty acids compete with glucose for insulin-mediated transport across the cell membrane.
D. fatty acids inhibit the formation of pyruvate from lactate.
E. fatty acids compete with glucose for NAD^+, which blocks the formation of pyruvate.

6. Which one of the following statements regarding gluconeogenesis is incorrect?

A. Glucose can be synthesized from non-carbohydrate precursors.
B. Gluconeogenesis is not a reversal of glycolysis.
C. Gluconeogenesis and glycolysis are reciprocally regulated.
D. Lactate and alanine formed by contracting muscle are converted into glucose in muscle.
E. Six high energy phosphate bonds are spent in synthesizing one molecule of glucose from pyruvate.

7. Which one of the following statements is NOT characteristic of gluconeogenesis?

A. It does not require energy in the form of ATP.
B. It is important in maintaining blood glucose during the normal overnight fast.
C. It involves the enzyme, fructose 1,6-bisphosphatase.
D. It uses carbon skeletons provided by degradation of amino acids.
E. It is not a reversal of glycolysis.

8. Which one of the following statements regarding glycogen metabolism is incorrect?

A. Glycogen consists of α-1,4-glycosidic bonds and α-1,6-glycosidic bonds.
B. Glycogen phosphorylase catalyzes the hydrolytic cleavage of glycogen into glucose-1-phosphate.
C. A debranching enzyme is needed for the complete breakdown of glycogen.
D. Phosphoglucomutase converts glucose-1-phosphate into glucose-6-phosphate.
E. Glycogen is synthesized and degraded by different pathways.

9. Epinephrine has which of the following effects on glycogen metabolism in the liver?

A. The net synthesis of glycogen is increased.
B. Glycogen phosphorylase is activated, whereas glycogen synthase is inactivated.
C. Both glycogen phosphorylase and glycogen synthase are activated.
D. Glycogen phosphorylase is inactivated, whereas glycogen synthase is activated.
E. Phosphoglucomutase is phosphorylated

10. There are three irreversible steps in glycolysis. They are:

A. hexokinase, phosphoglycerate kinase, and pyruvate kinase
B. hexokinase, phosphofructokinase, and pyruvate kinase
C. phosphofructokinase, aldolase, and phosphoglyceromutase
D. phosphoglucose isomerase, glyceraldehyde 3-phosphate dehydrogenase, and enolase
E. triose phosphate isomerase, phosphoglycerate kinase, and enolase

11. Which one of the following statements concerning the pentose phosphate pathway is true?

A. NADH is a major product of the oxidative branch.
B. The entire pathway is "off" when NADPH levels are high
C. It is not present in brain tissue.
D. When glucose 6-phosphate dehydrogenase is inhibited by one of its products, ribose 5-phosphate biosynthesis still occurs.
E. It provides an alternative pathway for the synthesis of glucose from glycerol.

12. If you dissolve a pure crystal of β-D-glucose in water and leave the solution for a long period of time, it is likely that the solution will contain:

A. the open-chain form of D-glucose, α-D-glucose, and β-D-glucose.
B. the open-chain form of D-glucose and β-D-glucose.
C. α-D-glucose and β-D-glucose.
D. α-D-glucose only.
E. β-D-glucose only.

13. An individual who cannot synthesize functional liver fructose 1,6-bisphosphatase would be primarily affected by:

A. a failure to resynthesize glucose from lactate produced during exercise
B. an inability to metabolize fructose
C. a lowered yield of ATP production per mole of glucose metabolized
D. a failure to split fructose diphosphates into triose phosphates
E. none of the above

14. Hydrolytic anemia associated with a deficiency of erythrocyte glucose-6-phosphate dehydrogenase is marked by:

A. increased accumulation of lipid hydroperoxides
B. exacerbation by drugs which are usually harmless
C. increased ratios of oxidized to reduced glutathione
D. a high ratio of NAD^+ to NADPH
E. all of the above are correct

15. The enzyme aldolase catalyzes:

A. formation of fructose-6-phosphate from glucose-6-phosphate
B. oxidation of the aldehyde group of glucose
C. oxidation of the aldehyde group of glyceraldehyde-3-phosphate
D. conversion of glyceraldehyde-3-phosphate to dihydroxyacetone-phosphate
E. formation of dihydroxyacetone phosphate and glyceraldehyde-3-phosphate from fructose-1,6-bisphosphate

16. Which one of the following cases represents a metabolically important control?

A. Substrate oxidation in mitochondria is enhanced by elevated ATP levels.
B. The activity of liver hexokinase is stimulated by high levels of its product, glucose 6-phosphate.
C. Acetyl-CoA inhibits pyruvate carboxylase in the liver.
D. Phosphofructokinase (PFK) is inhibited by high levels of its substrate, MgATP.
E. Glucagon stimulates hepatic glycogen synthase.

17. Galactosemia is caused by

A. deficiency of galactose-1-phosphate uridyltransferase
B. deficiency of UDP-galactose 4-epimerase
C. the high content of lactose in artificial feeding formulae for babies
D. absorption of non-hydrolyzed lactose through the intestinal mucosa
E. excessive conversion of glucose-1-phosphate into galactose-1-phosphate

18. During the period of no food intake between the evening meal and breakfast the next day, the brain:

A. derives very little of its energy from the oxidation of amino acids.
B. metabolizes its own stores of glycogen.
C. does not utilize fatty acids
D. uses blood glucose derived from breakdown of hepatic glycogen.
E. all of the above are correct

19. In glucose 6-phosphate dehydrogenase deficiency, increased red cell lysis is ultimately due to:

A. problems with ATP production in mitochondria
B. a deficiency in ability to carry out glycolysis
C. increased leakage of K ion into the cells
D. an intrinsic deficiency of membrane structure
E. inability of the cell to maintain normal concentrations of NADPH

20. Biotin is required as a coenzyme in which one of the following reactions?

A. α-ketoglutarate + NAD^+ + CoA $\longrightarrow$ succinyl CoA + CO_2 + NADH
B. pyruvate + CO_2 + ATP $\longrightarrow$ oxaloacetate + ADP + P_i
C. pyruvate + NAD^+ + CoA $\longrightarrow$ acetyl CoA + CO_2 + NADH
D. 6-phosphogluconate $\longrightarrow$ ribulose-5-phosphate + CO_2
E. α-ketoglutarate + CO_2 + NADH $\longrightarrow$ isocitrate + NAD^+

21. Enzymatic hydrolysis of starch (amylose) by amylase:

A. occurs in the stomach and small intestine
B. yields products that contain 1,6-linked glucose chains only
C. involves an enzyme released by the pituitary gland
D. Yields products that are absorbed in the large intestine.
E. occurs in the liver

MATCHING: For each set of questions, choose the ONE BEST answer from the list of lettered options above it. An answer may be used one or more times, or not at all.

Questions 22 - 26

A. Glucose-1-phosphate
B. Glucose-6-phosphate
C. Glucose-1,6-bisphosphate
D. UDP-glucose
E. None of the above

22. Activated form of glucose utilized by glycogen synthetase.

23. Generated during breakdown of glycogen by glycogen phosphorylase.

24. Coenzyme for phosphoglucomutase.

25. Generated by the hexokinase reaction.

26. Substrate for phosphohexose isomerase.

Questions 27 - 29:

A. glucose-6-phosphate
B. UDP-galactose
C. lactate
D. acetyl CoA
E. 1,3-bisphosphoglycerate

27. Regulates glycogen synthase.

28. Regulates pyruvate carboxylase.

29. Cannot be converted to glucose.

Questions 30 - 33:

A. glycerol → glucose
B. pyruvate → acetyl CoA
C. glucose → fatty acid
D. glucose → lactate
E. glucose → fructose

In mammalian metabolism the pathway that:

30. Requires the operation of the hexose monophosphate shunt.

31. Cannot be reversed to achieve net synthesis of glucose.

32. Occurs without phosphorylation of glucose.

33. Allows triglycerides to contribute to glucose synthesis.

Questions 34 - 36:

A. ATP is a substrate
B. ATP is an inhibitor
C. AMP is an inhibitor
D. ATP is both a substrate and an inhibitor

34. Phosphofructokinase.

35. glucokinase.

36. fructose-1,6-bisphosphatase.

Questions 37 - 40:

    A. glucose-6-P dehydrogenase deficiency
    B. glucose-6-phosphatase deficiency
    C. galactokinase deficiency
    D. UDP-galactose epimerase deficiency
    E. hexokinase deficiency

37. Increased concentration of galactose-1-P

38. Increased lipid peroxides in erythrocytes

39. Increased concentration of liver glycogen

40. Essentially benign excretion of a glucose epimer

Questions 41 - 43:

    A. galactose
    B. glucose
    C. fucose
    D. N-acetylgalactosamine
    E. glucosamine

41. Dextran

42. Lactose

43. Hyaluronic acid

> **DIRECTIONS:** For each of the following multiple-choice questions (44 - 100), choose the ONE BEST answer.

---

44. During starvation, as gluconeogenesis increases to maintain the levels of blood glucose, which one of the following will be enhanced?

A. liver pyruvate kinase activity
B. the secretion of insulin by the pancreas
C. muscle glucose-6-phosphatase activity
D. the metabolism of acetyl CoA to pyruvate
E. the metabolism of glutamate to glucose-6-phosphate

45. In hereditary fructose intolerance, the primary biochemical defect is:

A. a deficiency in activity of an aldolase isozyme
B. increased allosteric sensitivity of phosphofructokinase to AMP
C. inhibition of glycogen synthetase
D. an inability to absorb fructose
E. a deficiency in the activity of fructokinase

46. Hyaluronic acid is a:

A. glycoprotein
B. high molecular weight, positively charged polysaccharide
C. polymer which contains sulfate
D. repeating disaccharide of glucuronic acid and N-acetylglucosamine
E. lipoprotein

47. Glucose, labeled with ^{14}C in different carbon atoms is added to a tissue that is rich in the enzymes of the hexose monophosphate shunt. Which one will give the most rapid initial evolution of $^{14}CO_2$?

A. glucose-1-^{14}C
B. glucose-2-^{14}C
C. glucose-3,4-^{14}C
D. glucose-5-^{14}C
E. glucose-6-^{14}C

48. Glyceraldehyde-3-phosphate dehydrogenase produces which one of the following as a product?

A. inosine triphosphate
B. 1,3-bisphosphoglyceric acid
C. cytidine triphosphate
D. phosphoenolpyruvic acid
E. phosphocreatine

49 The tissue with the lowest activity for the oxidation of glucose-6-phosphate by the pentose phosphate pathway is:

A. liver
B. lactating mammary gland
C. striated muscle
D. adrenal cortex
E. adipose tissue

50. The absence of which one of the following reactions is responsible for the inability of man to use fatty acids in the *de novo* net synthesis of glucose?

A. oxaloacetate → pyruvate
B. oxaloacetate + acetyl CoA → citrate
C. acetyl CoA → pyruvate
D. pyruvate → phosphoenolpyruvate
E. phosphoenolpyruvate → oxaloacetate

51. Which one of the following substrates can NOT contribute to net gluconeogenesis in mammalian liver?

A. alanine
B. stearate
C. α-ketoglutarate
D. glutamate
E. pyruvate

52. Transketolase requires which one of the following coenzymes?

A. pyridoxal phosphate
B. lipoamide
C. thiamin pyrophosphate
D. cobalamin
E. tetrahydrofolic acid

53. Conversion of phosphorylase b to phosphorylase a involves a reaction with which amino acid in the enzyme's polypeptide backbone?

A. aspartic acid
B. serine
C. glycine
D. cysteine
E. arginine

54. Activated sugar residues utilized for the biosynthesis of complex glycoproteins include all EXCEPT:

A. GDP-mannose
B. dolichol phosphorylglucose
C. UDP-glucuronic acid
D. CDP-N-acetylneuraminic acid
E. UDP-galactose

55. Unusually high concentrations of liver glycogen could be produced by:

A. adrenocortical insufficiency
B. a lack of insulin
C. a lack of glycogen synthetase
D. alkalosis
E. a lack of phosphorylase kinase

56. In the metabolism of glycerol to glycogen, the first intermediate of glycolysis encountered is:

A. glyceraldehyde-3-phosphate
B. dihydroxyacetone phosphate
C. 3-phosphoglyceric acid
D. ribulose-5-phosphate
E. 1,3-bisphosphoglyceric acid

57. A normal person has received a large dose of insulin. Which one of the following events will occur?

A. increased epinephrine secretion
B. an increase in blood glucose
C. a decrease in blood glucose
D. decreased uptake of amino acids by muscle
E. increased blood glycogen synthesis

58. The increase of glycogenolysis in muscle produced by epinephrine may be attributed to:

A. decreased Ca^{++}
B. activation of aldolase
C. reduction in total NAD^+ plus NADH
D. conversion of phosphorylase b to phosphorylase a
E. conversion of glycogen synthase b to glycogen synthase a

59. Glycosaminoglycans are characterized by:

A. a branched chain
B. multiple cationic sites
C. ester sulfates
D. the presences of N-acetyl neuraminic acid
E. peptide bonds between disaccharide units

60. The major rate-limiting step of glycolysis is the:

A. conversion of glucose to glucose 6-phosphate
B. conversion of glucose 6-phosphate to fructose 6-phosphate
C. conversion of fructose 6-phosphate to fructose 1,6-bisphosphate
D. aldolase reaction
E. epimerase reaction

61. Glucagon:

A. has actions similar to those of insulin
B. is secreted by the beta cells of the pancreatic islets
C. increases cyclic AMP concentrations in many tissues
D. targets liver primarily
E. targets muscle primarily

62. The activity of glycogen phosphorylase in muscle is affected primarily by:

A. glucagon
B. melanocyte stimulating hormone
C. adrenocorticotropic hormone
D. epinephrine
E. somatostatin

63. Sialic acid is:

A. found only in mammalian tissues
B. the major carbohydrate found in heparin
C. a normal constituent of glycoproteins
D. an ϵ-carboxy amino acid
E. a cofactor for neuraminidase

64. Glycolysis in the red blood cell produces:

A. citric acid
B. NADH
C. GTP
D. CO_2
E. glucose-1-phosphate

65. In collagen a carbohydrate moiety is linked to:

A. threonine
B. hydroxyproline
C. hydroxylysine
D. asparagine
E. serine

66. Sucrose is:

A. a disaccharide containing glucose and fructose
B. a reducing disaccharide of plant origin
C. a disaccharide containing glucose and galactose
D. a fructose polymer
E. a product of digestion of cellulose

67. The major site of carbohydrate digestion is:

A. mouth
B. stomach
C. small intestine
D. large intestine
E. pancreas

68. Which one of the following is a ketose?

A. D-glucose
B. D-ribose
C. D-galactose
D. D-fructose
E. N-acetylglucosamine

69. Which one of the following is a gluconeogenic enzyme?

A. glucose-6-phosphate dehydrogenase
B. glucose-6-phosphatase
C. phosphofructokinase
D. pyruvate carboxylase
E. lactate dehydrogenase

70. Amylose is:

A. a branched homopolysaccharide
B. a linear homopolysaccharide
C. a linear heteropolysaccharide
D. a salivary enzyme
E. a pancreatic enzyme

71. Von Gierke's disease is characterized by massive enlargement of the liver, severe hypoglycemia, ketosis, hyperuricemia and hyperlipemia. It is caused by defective:

A. amylo-α-1,6 glucosidase
B. branching enzyme (α-1,4 $\rightarrow$ α-1,6)
C. glucose 6-phosphatase
D. α-1,4 glucosidase
E. phosphorylase

72. The catabolism of 1 mole of glucose in the glycolytic pathway under anaerobic conditions results in the formation (or NET gain) of:

A. 1 mole of UDP-galactose
B. 2 moles of ATP + 2 moles of lactic acid
C. 1 mole of glucose-1,6-diphosphate
D. 6 moles of CO_2 + 6 moles of H_2O
E. 1 mole of ethanol + 1 mole of lactic acid

73. The most significant role of the oxidative portion of the hexose monophosphate shunt is the generation of:

A. hexose monophosphate from free hexose
B. NADPH
C. NAD^+
D. ATP from ADP and P_i
E. UDP-gluconic acid from UDP-glucose

74. All of the following compounds are potential substrates (carbon sources) for the gluconeogenic pathway EXCEPT:

A. glutamic acid
B. aspartic acid
C. oleic acid
D. succinic acid
E. glycogen (from skeletal muscle)

75. A nine-carbon acid found as the non-reducing termini of oligosaccharide side chains of many glycoproteins is:

A. hyaluronic acid
B. sialic acid
C. iduronic acid
D. glucuronic acid
E. gluconic acid

76. Which one of the following carbohydrates contains a monosaccharide unit other than glucose?

A. Glycogen
B. Cellulose
C. Maltose
D. Lactose
E. Starch

77. Through which linkage are two sugar units connected in disaccharides?

A. O-glycosidic bond
B. Peptide bond
C. N-glycosidic bond
D. Phosphodiester bond
E. Hemiacetal

78. In mammals glucose 6-phosphate is converted to all of the following compounds EXCEPT which one?

A. glucose
B. fructose 1-phosphate
C. 6-phosphogluconolactone
D. fructose 6-phosphate
E. glucose 1-phosphate

79. A deficiency in hepatic phosphoglucomutase would most likely lead to a decrease of which one of the following cellular components:

A. pyruvate
B. ribose 5-phosphate
C. NADP
D. NADH
E. glycogen

80. Which one of the following is a glycolytic enzyme of liver and is activated by protein phosphate phosphatase in response to decreasing glucagon levels?

A. glycogen synthetase
B. glycogen phosphorylase
C. pyruvate kinase
D. triose phosphate isomerase
E. lactate dehydrogenase

81. Glyceraldehyde-3-phosphate is an intermediate in:

A. glycolysis
B. hexose monophosphate shunt
C. alcoholic fermentation
D. all of the above
E. none of the above

82. Glucagon promotes gluconeogenesis by stimulating the synthesis of:

A. cyclic AMP
B. ATP
C. 5'-AMP
D. glucose 6-phosphate
E. ribose 5-phosphate

83. An allosteric inhibitor of phosphofructokinase is:

A. α-ketoglutarate
B. oxaloacetate
C. citrate
D. isocitrate
E. succinate

84 The formation of which one of the following compounds renders the entry of glucose into most cells almost irreversible?

A. glucose 1-phosphate
B. pyruvate
C. PEP (phosphoenolpyruvate)
D. 6-phosphogluconate
E. glucose 6-phosphate

85. Which one of the following enzymes is inhibited by an accumulation of NADH?

A. aldolase
B. enolase
C. pyruvate kinase
D. glyceraldehyde 3-phosphate dehydrogenase
E. pyruvate decarboxylase

86. Conditions which favor gluconeogenesis include:

A. high citrate concentration
B. low acetyl-CoA concentration
C. high ADP concentration
D. high glucagon concentration
E. low fructose 1,6-bisphosphate concentration

87. The relative importance of the pentose phosphate pathway for glucose metabolism is:

A. equal among all tissues
B. greatest in those tissues involved in lipid biosynthesis
C. greatest in skeletal muscle
D. greatest in adult life
E. lowest in adipose tissue

88. Which one of the following glycogen storage disease does NOT involve a defect in the glycogen degradation pathway?

A. Type I (von Gierke's Disease)
B. Type IV (Andersen's Disease)
C. Type VI (Hers' Disease)
D. Type V (McArdle's Disease)
E. Type III (Cori's Disease)

89. Which one of the following sugars is found exclusively in glycosaminoglycans?

A. N-acetylglucosamine
B. N-acetylgalactosamine
C. sialic acid
D. L-iduronic acid
E. L-arabinose

90 Which one of the following enzymes is found mainly in liver or kidney cells?

A. glucose 6-phosphatase
B. hexokinase
C. phosphoglucoisomerase
D. pyruvate kinase
E. glycogen synthase

91. Which one of the following is the first enzyme unique to the pentose phosphate pathway?

A. lactonase
B. 6-phosphogluconate dehydrogenase
C. transaldolase
D. glucose 6-phosphate dehydrogenase
E. phosphoglucoisomerase

92. In a patient suffering from Type III glycogen storage disease, an abnormal glycogen exhibiting short outer branches is observed. Which enzyme is most likely to be defective?

A. glycogen synthase
B. amylo-1,6-glucosidase (debranching enzyme)
C. glycogen phosphorylase
D. phosphoglucomutase
E. UDP-glucose pyrophosphorylase

93. The number of residues bound by glycosidic linkages to a glucose residue that forms a branch point in glycogen is:

A. 1
B. 2
C. 3
D. 4
E. 5

94. What is the NET yield of NADH when glucose 6-phosphate is oxidized by anaerobic glycolysis to yield lactate?

A. 0
B. 1
C. 2
D. 3
E. 4

95. What is the NET yield of ATP when glucose 1-phosphate is oxidized by anaerobic glycolysis to lactate?

A. 0
B. 1
C. 2
D. 3
E. 4

96. The first step in the gluconeogenic pathway (starting with pyruvate) results in the formation of:

A. phosphoenolpyruvate
B. malate
C. aspartate
D. oxaloacetate
E. lactate

97. The first step of glycolysis catalyzed by hexokinase converts glucose to glucose 6-phosphate. This reaction may be classified as:

A. phosphoryl shift
B. isomerization
C. phosphoryl transfer
D. dehydration
E. aldol cleavage

98. All of the following are high energy compounds EXCEPT:

A. phosphoenolpyruvate
B. acetyl-CoA
C. glucose 6-phosphate
D. acetyl phosphate
E. creatine phosphate

99. All of the following statements about phosphofructokinase (PFK) are true EXCEPT:

A. it is a major control enzyme in glycolysis
B. ATP is a substrate for PFK
C. AMP is a negative effector of PFK
D. ATP is a negative effector of PFK
E. it catalyzes a metabolically irreversible reaction; ie., its equilibrium point lies far in one direction.

100. Epinephrine (in muscle) and glucagon (in liver):

A. activate adenyl cyclase
B. inactivate phosphorylase and activate glycogen synthetase
C. stimulate triglyceride synthesis
D. stimulate glycogen synthesis
E. act synergistically with insulin

XIII.   ANSWERS TO QUESTIONS ON CARBOHYDRATES

1.	D	21.	A	41.	B	61.	D	81.	D
2.	D	22.	D	42.	A	62.	D	82.	A
3.	B	23.	A	43.	D	63.	C	83.	C
4.	B	24.	C	44.	E	64.	B	84.	E
5.	D	25.	B	45.	A	65.	C	85.	D
6.	D	26.	B	46.	D	66.	A	86.	A
7.	A	27.	A	47.	A	67.	C	87.	B
8.	B	28.	D	48.	B	68.	D	88.	B
9.	B	29.	D	49.	C	69.	D	89.	D
10.	B	30.	C	50.	C	70.	B	90.	A
11.	D	31.	B	51.	B	71.	C	91.	D
12.	A	32.	B	52.	C	72.	B	92.	B
13.	A	33.	A	53.	B	73.	B	93.	C
14.	E	34.	D	54.	C	74.	C	94.	A
15.	E	35.	A	55.	E	75.	B	95.	D
16.	D	36.	C	56.	A	76.	D	96.	D
17.	A	37.	D	57.	C	77.	A	97.	C
18.	E	38.	A	58.	D	78.	B	98.	C
19.	E	39.	B	59.	C	79.	E	99.	C
20.	B	40.	D	60.	C	80.	C	100.	A

4. ENERGETICS AND BIOLOGICAL OXIDATION

Thomas Briggs

I. CONCEPTS IN BIOLOGICAL OXIDATION

A. Oxidation and Reduction

Biological oxidation provides most of the energy for aerobic metabolism. This energy is released when electrons are transferred from fuel molecules to oxygen. Biological oxidation effects this transfer in a controlled way, and conserves much of the energy in the form of phosphoanhydride bonds of ATP, a useful molecule which can then supply the energy to drive a multitude of energy-consuming processes. The process of phosphorylation of ADP to ATP, driven by the transfer of electrons to oxygen, is called **oxidative phosphorylation**.

Oxidation can be defined in three ways:
1. addition of oxygen;
2. removal of hydrogen;
3. **removal of electrons**: this is the most general definition.

Reduction is the converse of the above.

Biological oxidation and reduction are always linked. For every oxidation, there *must* be a reduction; for every electron donor, there *must* be an electron acceptor.

B. Thermodynamics

For a reaction $A + B \rightleftharpoons C + D$ at equilibrium, the **equilibrium constant**, K_{eq}, is given by the expression:

$$K_{eq} = \frac{[C][D]}{[A][B]}$$

Any reaction proceeds with a change in **free energy**, the useful energy produced, the energy available to do useful work. The observed change is denoted by ΔG, (sometimes incorrectly called ΔF). ΔG depends on the **nature** of the reaction and the **concentrations** of reactants and products, and has predictive power: a **negative** ΔG means the reaction will go to the **right** as written, or down an energy hill. The reaction is said to be "spontaneous." A **positive** ΔG means the reaction will go to the **left**. *At equilibrium, $\Delta G = 0$.*

Note that ΔG *tells nothing about rates of reactions.* A "spontaneous" process may not proceed by itself at a measurable speed, due to the presence of an activation energy barrier. One function of enzymes is to lower this barrier and speed up the rate. The final position of equilibrium is related to ΔG and **not** affected by enzymes. Thus an enzyme will not affect the ΔG of a reaction.

If a reaction starts at **standard conditions** of pH 7 and 1 M concentrations of all reactants and products, and goes to equilibrium, then the change in free energy is the **standard** free energy change (denoted by superscript zero and prime) and is related to K_{eq} (R is the gas constant; T, the absolute temp.):

$$\Delta G^{\circ\prime} = -RT \ln K_{eq}$$

A reaction with an equilibrium constant >1 will have a **negative** $\Delta G^{\circ\prime}$, and will *tend* to go to the right. But this tendency may be reversed by **concentration** effects or input of **energy** from somewhere else. Thus a reaction can be driven against an unfavorable equilibrium by mass action, or if it is coupled, through a common intermediate, with an energy-releasing reaction.

When reactions are coupled, or occur in series, ΔG's of component reactions are simply added. It is the overall ΔG, the algebraic sum of individual ΔG's, that determines whether the process as a whole will occur spontaneously.

The degree of **disorder**, or randomness, is called **entropy** (S). ΔS is the entropy change associated with a particular transformation. Entropy is higher in a less-ordered system, such as a denatured protein. Entropy of living systems is kept low (highly ordered) at the expense of an increase in entropy of the surroundings. A spontaneous reaction **can** proceed with a decrease in entropy, if the **total** entropy change, including that of surroundings, is positive.

In an oxidation-reduction reaction, reducing power, E, also called **redox potential**, is the capacity to donate electrons. It can be measured, in volts, by comparing a component of the reaction with a standard hydrogen electrode.

$\Delta E_o'$ is the difference in **standard** redox potential (standard conditions) between electron donor and acceptor. The component with the more negative E_o' is the stronger electron donor or reducing agent, and will tend to reduce (be oxidized by) the second component. A large K_{eq} (i.e., reaction tends to proceed far to the right) is associated with a large **positive** $\Delta E_o'$. Thus $\Delta E_o'$ is mathematically related to $\Delta G^{o'}$ but the two are opposite in sign.

Points to remember:

1. A process that goes "spontaneously" has a **negative** ΔG.

2. An oxidation-reduction reaction that goes "spontaneously" has a **positive** ΔE.

3. A substance with the more **negative** redox potential will reduce another with a less negative or more positive redox potential: *electrons tend to flow from* **E negative to E positive**.

4. The **molecular oxygen** system, toward which electrons flow in biological oxidation, has the **most positive** redox potential.

5. The directional tendency that a reaction might normally have *may* be reversed by:
   a. a sufficient difference in **concentration** between reactants and products (mass action);
   b. input of sufficient **energy**.

C. High-Energy Compounds

When ATP, an anhydride of phosphoric acid, undergoes hydrolysis and the terminal phosphate is transferred to H_2O, the equilibrium lies far to the right and an unusually large amount of free energy is released:

$$\text{ATP} + H_2O \rightleftharpoons \text{ADP} + P_i \qquad \Delta G^{o'} = -7.3 \text{ Kcal/mole}$$

This is in part because the products of the reaction are stabilized by a greater number of resonance forms. The reaction can be reversed by coupling it to another reaction that supplies an equal or greater amount of energy. ATP is said to be a "high-energy" compound.

Substances such as ATP serve an "energy-currency" function by accepting energy from electron transfer and the metabolism of certain substrates, and in turn supplying it to drive various energy-consuming functions: synthesis, muscle contraction, the maintenance of ionic gradients, etc. Table 4-1 shows representative high-energy compounds. Note that ordinary esters of phosphoric acid, such as glucose-6-phosphate, are *not* high-energy compounds.

Compound	Type	$\Delta G°'$, Kcal/mole
Phosphoenolpyruvate	Anhydride of enol and phosphoric acid	-14.8
1,3-Diphosphoglycerate	Anhydride of carboxylic and phosphoric acids	-11.8
Phosphocreatine	Guanidino phosphate	-10.3
Acetyl phosphate	Anhydride of carboxylic and phosphoric acids	-10.1
Acetyl CoA	Thioester	-7.5
ATP	Anhydride of phosphoric acid	-7.3
Glucose-6-phosphate	Ester of phosphoric acid	-3.3
Glycerol-1-phosphate	Ester of phosphoric acid	-2.2

Table 4-1. Standard Free Energy of Hydrolysis of Some Phosphorylated Compounds.

II. METABOLISM OF PYRUVATE TO CARBON DIOXIDE

A. The Pyruvate Dehydrogenase Complex

In aerobic metabolism, the major fate of the pyruvate produced by glycolysis is transport into mitochondria, then conversion to acetyl CoA and CO_2 by the *pyruvate dehydrogenase complex*, which carries out the over-all reaction:

$$\text{Pyruvate} + \text{CoASH} + \text{NAD}^+ \longrightarrow \text{Acetyl CoA} + CO_2 + \text{NADH} + \text{H}^+$$

The reaction is, for practical purposes, irreversible. The complex (from *E. coli*) contains multiple copies of several types of subunits: 24 of *pyruvate decarboxylase* (E1), 24 of *dihydrolipoyl transacetylase* (E2), and 12 of *dihydrolipoyl dehydrogenase* (E3). Five cofactors participate in the reaction. Three are bound to components of the complex: thiamin pyrophosphate (TPP), lipoic acid (Lip), and flavin adenine dinucleotide (FAD); two come and go from solution: nicotinamide adenine dinucleotide (NAD^+) and coenzyme A (CoA). The different subunits act in a coordinated fashion as follows (Figure 4-1):

1. Through the carbonyl carbon, pyruvate binds to TPP and loses CO_2, leaving hydroxyethyl TPP on E1.

2. Hydroxyethyl is oxidized to an acetyl group as it is transferred to lipoamide on E2. The resultant thioester is energy-rich.

3. The acetyl group is transferred to CoA and leaves as acetyl CoA. Lipoamide is left with both thiols in the reduced state.

4. Lipoamide on E2 is reoxidized by FAD on E3.

5. Reduced $FADH_2$ on E3 is reoxidized by NAD^+, leaving FAD on E3. The entire complex is back in its original state; two H's from the oxidation of pyruvate are carried away as $\text{NADH} + \text{H}^+$.

Regulation. The pyruvate dehydrogenase complex is subject to end-product inhibition, i.e., NADH and acetyl CoA inhibit the enzyme. In addition, the complex from eukaryotic sources contains a kinase which can phosphorylate a serine residue in the presence of ATP; the phosphorylated enzyme has decreased activity. A phosphatase, also present in the complex, restores activity. Thus the supply of ATP, signifying the energy-state of the cell, leads to regulation of this important enzyme in a major energy-producing pathway.

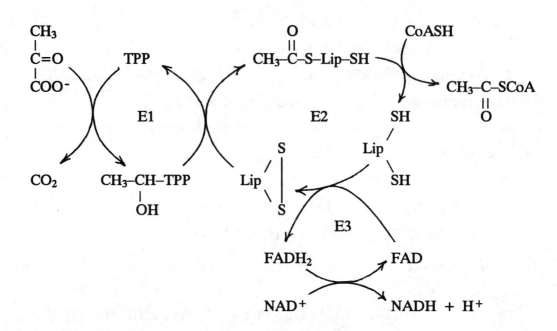

Figure 4-1. Reactions Catalyzed by the *Pyruvate Dehydrogenase Complex.*

    Enzyme 1 (E1): *Pyruvate decarboxylase*
    Enzyme 2 (E2): *Dihydrolipoyl transacetylase*
    Enzyme 3 (E3): *Dihydrolipoyl dehydrogenase*
    Cofactors: Thiamin pyrophosphate (TPP)
              Lipoamide (Lip)   Coenzyme A (CoA)
              NAD^+ FAD

B. The Citric Acid Cycle (Krebs Tricarboxylic Acid Cycle)

The cycle is only active under aerobic conditions. It metabolizes the remaining carbons of glucose, as well as all carbons of fatty acids, to CO_2. Hydrogens are collected on carriers (NADH, $FADH_2$) for subsequent oxidation by the electron transport system. The cycle also serves to link many metabolic pathways not only in catabolism, but also in an anabolic mode, since it can provide intermediates for gluconeogenesis, synthesis of amino acids, etc.

It functions as a closed loop, in a sense catalytically, in that each turn regenerates the starting material (oxaloacetate). Products of the cycle are two CO_2, one high-energy phosphate, and four *reducing equivalents* (3 NADH and 1 $FADH_2$). Although some individual steps are reversible, the cycle contains enough steps with large negative changes in free energy that as a whole it is not reversible. Nearly all the enzymes are soluble, occurring in the mitochondrial matrix.

The cycle operates as follows (Figure 4-2):

1. *Condensation.* The two-carbon moiety from acetyl CoA condenses with oxaloacetate producing the six-carbon tricarboxylic acid, citrate (***citrate synthase***).

2. *Isomerization.* Citrate is dehydrated to *cis*-aconitate, then rehydrated to isocitrate (***aconitase***).

3. *Oxidative decarboxylation.* Isocitrate is converted to α-ketoglutarate with loss of CO_2. This carbon and that of the CO_2 in the subsequent step are derived from the oxaloacetate in step 1, not from the incoming acetyl group. Two H's are released as $NADH + H^+$ (***isocitrate dehydrogenase***).

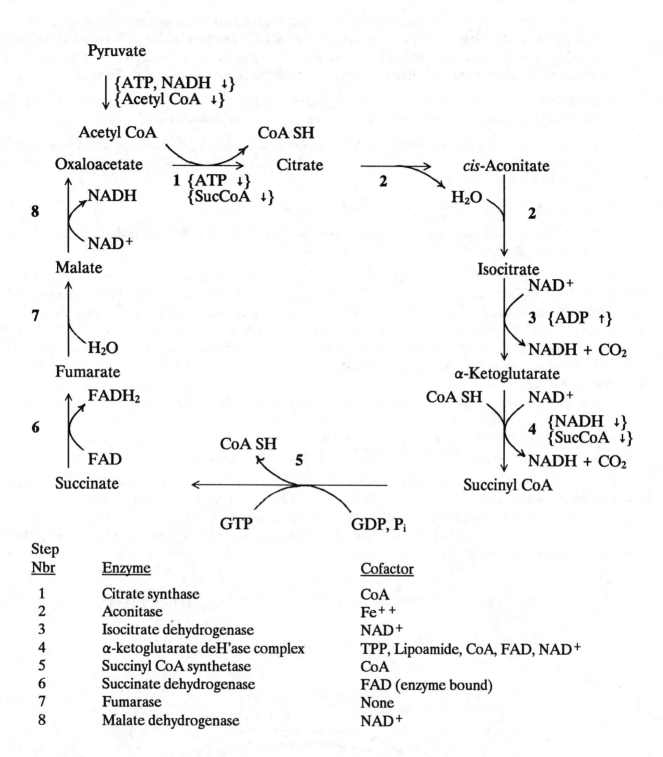

Figure 4-2. The Citric Acid Cycle.

Numbers in boldface refer to enzymes.

Items in curved brackets refer to regulatory effects: ↑ = stimulatory; ↓ = inhibitory.

4. A second *oxidative decarboxylation* occurs as α-ketoglutarate is converted to a four-carbon acid (succinate, as the CoA derivative), again with loss of CO_2 and production of a reducing equivalent ($NADH + H^+$) (*α-ketoglutarate dehydrogenase*). The enzyme is a complex whose constitution and mechanism of action are very similar to pyruvate dehydrogenase and its mode of action.

5. *Phosphorylation*. The high-energy of succinyl CoA is conserved during the generation of GTP from GDP and P_i (*succinyl CoA synthetase*, formerly known as *succinyl thiokinase*).

6. *Oxidation*. Two H's are removed (as $FADH_2$) from succinate to form the unsaturated acid fumarate (*succinate dehydrogenase*). It often happens that FAD is the electron acceptor when two H's are removed from two adjacent carbon atoms.

7. *Hydration*. Water adds to fumarate to form the hydroxy acid, malate (*fumarase*).

8. *Oxidation*. Oxaloacetate is regenerated as two final H's are carried off from malate as $NADH + H^+$. (*malate dehydrogenase*).

Regulation. The citric acid cycle is sensitive to the supply of substrates at several points. *Citrate synthase* requires both acetyl CoA and oxaloacetate; if the supply of the former decreases (due to slowed glycolysis or decreased fatty acid oxidation), or if oxaloacetate is diverted for gluconeogenesis, this step and the cycle as a whole will slow down. All four oxidation-reduction reactions need oxidized coenzymes (NAD^+, FAD) in order to function. Thus the cycle is also responsive to the redox state of the cell, and will not operate unless oxidized coenzymes are regenerated by the electron transport system. In addition, there is control by effectors at several points:

     a. *citrate synthase* is allosterically inhibited by ATP and also by succinyl CoA;

     b. *isocitrate dehydrogenase* is allosterically activated by ADP;

     c. the α-ketoglutarate dehydrogenase complex is controlled by end-product inhibition very much as is pyruvate dehydrogenase: it is slowed by NADH and, in this case, succinyl CoA.

Control of the pyruvate dehydrogenase complex and of the citric acid cycle illustrates the application of the concept of "energy charge." In analogy with a storage battery, a cell is said to have a high charge if the amount of ATP, as a fraction of the total amount of adenine nucleotides, is high. In the regulatory functions discussed above, a high "energy charge" inhibits those processes that result in production of energy; a state of low "charge" stimulates energy-producing processes.

III. ELECTRON TRANSFER VIA THE RESPIRATORY CHAIN

The Mitochondrion: about 1 x 3 μm (Figure 4-3). A liver cell has about 1000. The inner membrane, of

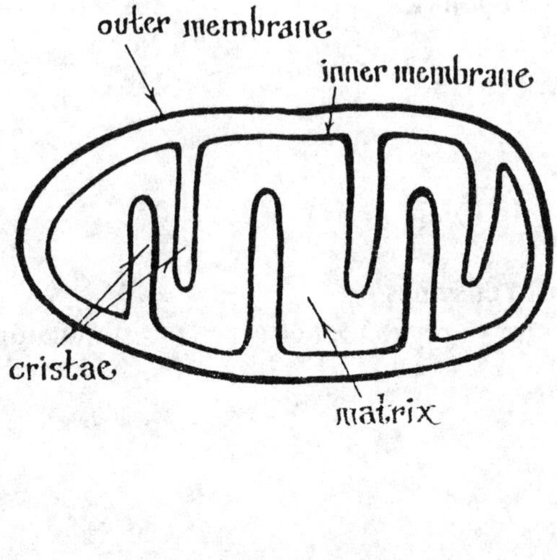

Figure 4-3.
The Mitochondrion.

which cristae are extensions, is *very selective in its permeability*. Among its many enzymes are those of electron transport and succinate dehydrogenase. The matrix has the other enzymes of the citric acid cycle, pyruvate dehydrogenase, and many others.

In catabolism, the usual electron (and H) acceptor is NAD^+. Flavoproteins are also used: protein-bound FAD, or FMN. Reduced coenzymes ($FMNH_2$, $FADH_2$, $NADH + H^+$) *must be reoxidized*. In aerobic metabolism, this is by the **electron transport system**.

Electron Transport System (ETS): a chain of enzymes, arranged in three complexes located on the **inner mitochondrial membrane**, specialized to carry out electron transfer from reduced coenzymes to oxygen. At each transfer, a drop in free energy (ΔG is negative) occurs. At each of the three complexes, sufficient free energy is released ultimately to drive the phosphorylation of ADP to ATP (Figure 4-4).

Several enzymes of the ETS are **cytochromes** — heme proteins. In the cytochrome system, electrons are transferred because the valence of the iron can change from $Fe^{++} \rightleftharpoons Fe^{+++}$:

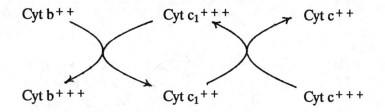

An electron is transferred from Cyt b to Cyt c_1; b is oxidized and c_1 is reduced.   Now the electron passes from reduced Cyt c_1 to Cyt c. c_1 is reoxidized; c is reduced.

Coupled Oxidative Phosphorylation: the process of ATP generation is tightly **coupled** to the process of electron transfer. If one stops, the other stops, like gears that mesh. This is the normal state in respiring mitochondria.

Uncoupling: some chemicals, e.g. 2,4-dinitrophenol (DNP), can uncouple phosphorylation, allowing electron transfer to oxygen to proceed. Energy is wasted as heat; no ATP is formed. Like depressing the clutch on a car, the engine runs but no useful work is done.

P/O Ratio: the number of ATP's formed per atom of O consumed in metabolism of substrate. NADH has P/O of 3; $FADH_2$, only 2 because Complex 1 is bypassed (see Figure 4-4). An uncoupled system has P/O of zero for all substrates.

Regulation depends on "energy charge," of which the **ATP/ADP ratio** is in part an indicator. If ATP predominates (high energy charge), the cell doesn't need energy, and electron transfer slows. If the cell needs energy, ADP predominates and is available for coupled phosphorylation, which now speeds up, allowing increased electron transfer, re-oxidation of NADH and $FADH_2$, and speeding up of Krebs (TCA) cycle. A supply of the **reduced coenzymes** and of P_i are also necessary. Availability of O_2, the terminal electron acceptor, also limits, since lack of an acceptor would stop everything (Figure 4-4).

Yield of ATP: The complete aerobic metabolism of glucose to CO_2 and H_2O produces 38 ATP. The origin of the reducing equivalents and resulting high-energy phosphates is summarized in Table 4-2.

Electron Transport System	Comments

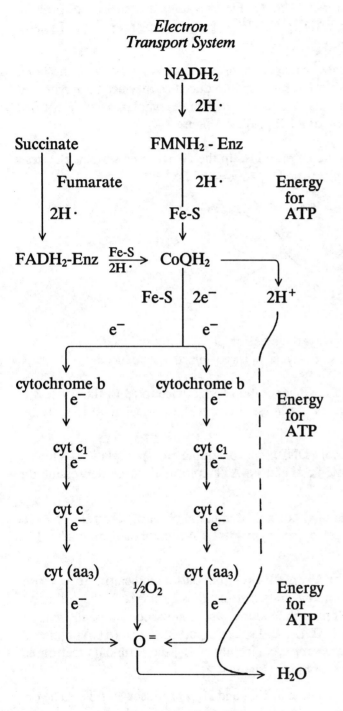

Electron transfer from NADH to CoQ mediated by Complex 1: *NADH dehydrogenase*. Inhibited by **rotenone**.

Energy is conserved here for generation of ATP while a pair of electrons (with 2H's) is transferred. This is the **first of 3 complexes** that produce energy for coupled oxidative phosphorylation.

Coenzyme Q is a quinone (ubiquinone). Electrons from succinate enter ETS here. Since the first energy-producing complex is bypassed, $FADH_2$ from succinate can produce only 2 ATP's.

Electron transfer from CoQ to Cyt c is mediated by *Cyt c Reductase*, the **2nd energy-producing complex**. Protons and electrons now separate, but it still takes a **pair** of electrons to generate enough energy for 1 ATP. **Antimycin A** inhibits here.

All cytochromes contain a protein plus heme (i.e., iron protoporphyrin) or a heme derivative.

Cytochrome (aa_3) is also called *cytochrome oxidase*. Contains Cu as well as Fe. This is the **3rd energy-producing complex**. Inhibited by **cyanide** (CN^-).

Energy for ATP: e^- transfer expels H^+ from mitochondrion. Re-entry drives phosphorylation by **chemi-osmotic coupling**.

Figure 4-4. Mitochondrial Electron Transport.

Transformation	Enzyme	Reducing Equivalent Produced	~P Produced	
			Substrate Level	Electron Transport Level
Glucose → → 1,3-Diphospho-glycerate	Glyceraldehyde-3-phosphate dehydrogenase	NADH + H$^+$ (x 2)		6 ATP
1,3-Diphospho-glycerate → 3-phosphoglycerate	Phosphoglycerate kinase		(2 ATP)*	
Phosphoenolpyruvate → pyruvate	Pyruvate kinase		2 ATP	
Pyruvate → acetyl CoA	Pyruvate dehydrogenase	NADH + H$^+$ (x 2)		6 ATP
Isocitrate → α-ketoglutarate	Isocitrate dehydrogenase	NADH + H$^+$ (x 2)		6 ATP
α-Ketoglutarate → succinyl CoA	α-Ketoglutarate dehydrogenase complex	NADH + H$^+$ (x 2)		6 ATP
Succinyl CoA → succinate	Succinyl CoA synthase		2 ATP (via GTP)	
Succinate → fumarate	Succinate dehydrogenase	FADH$_2$ (x 2)		4 ATP
Malate → oxaloacetate	Malate dehydrogenase	NADH + H$^+$ (x 2)		6 ATP

* 2 ATP's are consumed at start of glycolysis.   4 ATP   34 ATP

Assumptions: (1) tightly-coupled system;

(2) no loss of energy in transferring NADH from out-side of mitochondrion (glycolysis) to inside. Actually, yield may be reduced, depending on which "shuttle" mechanism is used.

Total: 38 ATP
from each glucose

Table 4-2. Yield of Reduced Coenzymes and ATP from the Complete Oxidation of Glucose

IV. CHEMI-OSMOTIC THEORY OF OXIDATIVE PHOSPHORYLATION

The components of the electron transport system are asymmetrically placed in the inner mitochondrial membrane, such that electron transfer results in a **directional extrusion of H$^+$** (Figure 4-5) from inside (matrix) to the intermembrane space. The outer membrane is porous and highly permeable to protons, making the intermembrane space equivalent to the cytoplasm. Since the inner membrane is impermeable to protons (except through the F_O - F_1 ATPase) this leads to a **pH difference** (more acid outside) and an **electrochemical gradient** across the inner membrane, a condition of potential energy. To release the potential energy, protons are allowed back in through the F_O - F_1 ATPase in such a way (details unclear) as to drive the phosphorylation of ADP to ATP.

A coupled system is self-regulated by high energy charge: unavailability of ADP prevents further phosphorylation which, in turn, prevents entry of H$^+$ through the ATPase. As the electrochemical gradient builds up to a maximal level, further extrusion of protons is inhibited, slowing electron transfer. Uncoupling agents act by conducting protons across the membrane so as to bypass the ATPase and "short-circuit" the gradient, allowing electron transfer without phosphorylation.

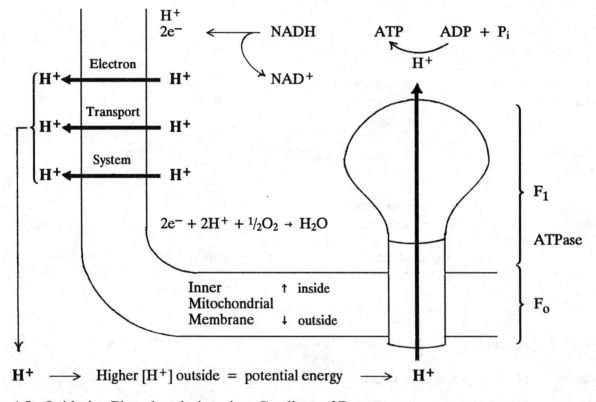

Figure 4-5. Oxidative Phosphorylation via a Gradient of Protons

V. REVIEW QUESTIONS ON ENERGETICS AND BIOLOGICAL OXIDATION

> ***DIRECTIONS:*** For each of the following multiple-choice questions (1 - 47), choose the ONE BEST answer.

1. In the reaction below, which of the following is true?

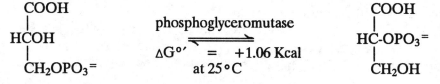

A. The K_M can be calculated from the amounts of substrates at equilibrium.
B. The reaction can occur only in aerobic metabolism.
C. The reaction goes faster at 20°C than at 25°C.
D. At equilibrium the substrate on the left has a higher concentration than the one on the right.
E. The presence of increased amount of enzyme increases the value of $\Delta G°'$.

2. Concerning entropy, which of the following is (are) correct?

A. Entropy means the degree of disorder or randomness.
B. Living things maintain low entropy by producing an increase in entropy of the surroundings.
C. Denatured macromolecules have higher entropy than the native forms.
D. A reaction may be spontaneous even if its change in entropy is negative.
E. All of the above are correct.

3. Which of the following is NOT correct?

A. A reaction is at equilibrium when the free energy change is zero.
B. A reaction may proceed spontaneously if the free energy change is negative.
C. A reaction which would otherwise not proceed can be made to proceed if it is coupled with another reaction for which the free energy change is negative.
D. The free energy change corresponds to (is proportional to) the change in the heat content of the system.
E. The concentrations of reactants and products are not necessarily equal at equilibrium.

4. Which of the following is NOT involved in mitochondrial electron transport?

A. NADH
B. cytochrome P_{450}
C. nonheme iron
D. cytochrome b
E. flavoprotein

5. The complete metabolism of one mole of pyruvate to CO_2 and H_2O produces about how many moles of ATP?

A. 12
B. 15
C. 30
D. 38
E. 60

6. In mitochondrial electron transport, the link between flavoprotein and the cytochrome system is:

A. ferritin
B. NADH - cyt c reductase
C. antimycin A
D. vitamin K
E. CoQ (ubiquinone).

7. Which of the following is necessary for a reaction to proceed spontaneously?

A. $\Delta G = 0$
B. $\Delta G < 0$
C. $\Delta G > 0$
D. $\Delta S > 0$
E. $\Delta H < 0$

8. One reaction can easily be coupled to another if

A. the two reactions have a common intermediate
B. one reaction has a more negative $\Delta G^{\circ\prime}$ than the other
C. one reaction has a more positive $\Delta G^{\circ\prime}$ than the other
D. the reactions are of the same type
E. one reaction has a greater rate than the other.

9. In a tightly-coupled system, the reaction or sequence producing the most ATP is:

A. isocitrate to malate
B. acetate to CO_2 and H_2O
C. succinate to CO_2 and H_2O
D. succinate to oxaloacetate
E. succinate to fumarate.

10. Of the following phosphate compounds, which has a more negative $\Delta G^{\circ\prime}$ of hydrolysis than GTP?

A. glucose-1-phosphate
B. glucose-6-phosphate
C. 2,3-diphosphoglycerate
D. phosphoenolpyruvate
E. glycerol-1-phosphate

11. Creatine phosphate

A. is an intermediate in arginine biosynthesis
B. is a specific inhibitor of aspartate transcarbamoylase
C. can serve as an immediate source of energy by direct interaction with actomyosin
D. is an allosteric effector for serine esterase
E. is a "high-energy" compound containing a P-N bond.

12. Which of the following biological processes results in a net increase in the chemical energy of the system?

A. CO_2 fixation
B. anaerobic glycolysis
C. photosynthesis
D. protein synthesis
E. aerobic phosphorylation

13. The condensation of oxaloacetate with acetyl CoA has a standard free energy change of –7.7 Kcal/mole. The equilibrium constant for the reaction

A. is greater than one
B. is less than one
C. depends on the concentrations of reactants and products
D. depends upon a catalyst
E. cannot be determined from the data given.

14. The most ATP per gram is yielded by which substrate?

A. isocitric acid
B. aspartic acid
C. oleic acid
D. fructose
E. glycogen

15. The equilibrium constant for the reaction:

malate + $NAD^+ \rightleftharpoons$ oxaloacetate + NADH + H^+

is defined by:

A. $K = \dfrac{(oxaloacetate)(NAD^+)(H^+)}{(malate)(NADH)}$

B. $\dfrac{1}{K} = \dfrac{(oxaloacetate)(NADH)(H^+)}{(malate)(NAD^+)}$

C. $K = \dfrac{(oxaloacetate)(NADH)(H^+)}{(malate)(NAD^+)}$

D. $K = \dfrac{(malate)(NAD^+)}{(oxaloacetate)(NADH)(H^+)}$

E. $K = \dfrac{(malate)(NAD^+)(H^+)}{(oxaloacetate)(NADH)}$

16. About how many grams of tristearin will yield the same amount of energy as that obtained from the metabolism of 100 grams of glucose?

A. 20
B. 40
C. 80
D. 100
E. 180

17. All of the following are "high-energy" compounds EXCEPT:

A. phosphocreatine
B. phosphoenolpyruvate
C. AMP
D. GTP
E. ATP.

18. Which of the following is always involved in biological oxidation-reduction reactions?

A. transfer of hydrogens
B. formation of water
C. mitochondria
D. transfer of electrons
E. direct participation of oxygen

19. If the hydrolysis of glucose-6-phosphate has K_{eq} = 100, and the phosphorylation of glucose by ATP to form glucose-6-phosphate has K_{eq} = 1000, then the hydrolysis of ATP to ADP and P_i has K_{eq} =

A. 10
B. 1×10^{-1}
C. 1×10^{3}
D. 1×10^{5}
E. 1×10^{-5}.

20. In humans, which of the following enzyme-catalyzed reactions does NOT produce CO_2?

A. isocitrate dehydrogenase
B. pyruvate dehydrogenase
C. α-ketoglutarate dehydrogenase
D. 6-phosphogluconate dehydrogenase
E. succinate dehydrogenase

21. If a preparation of healthy mitochondria is incubated with excess succinate, which of the following will stimulate oxygen uptake the most?

A. glucose-1-phosphate
B. oligomycin
C. malonate
D. ATP
E. ADP + P_i

22. When NADH is transformed to NAD^+, it loses

A. one hydronium ion
B. one hydride ion
C. one electron
D. two electrons
E. two protons and one electron.

23. Uniformly labeled ^{14}C-oxaloacetate is condensed with unlabeled acetyl CoA. After a single turn around the tricarboxylic acid cycle back to oxaloacetate, what fraction of the original radioactivity will be found in the oxaloacetate?

A. all
B. $3/4$
C. $1/2$
D. $1/4$
E. $1/3$

24. The complete oxidation of one mole of glucose in a biological system leads to the formation of about how many moles of ATP from ADP and P_i?

A. 2
B. 7
C. 14
D. 35
E. 70

25. Mitochondria

A. carry out glycolysis
B. make lipoprotein
C. conduct oxidative phosphorylation
D. synthesize sterols
E. hydroxylate drugs.

26. In mitochondrial electron transfer from cytochrome b to cytochrome c:

A. cytochrome c Fe^{+++} is reduced to cytochrome c Fe^{++}
B. it occurs because cyt c is a stronger oxidizing agent than cyt b
C. enough energy is produced for eventual phosphorylation of one ADP to ATP
D. cytochrome b Fe^{++} is oxidized to cytochrome b Fe^{+++}
E. all of the above are correct.

27. How many high-energy phosphates can result from the over-all reaction:

isocitrate + $^3/_2 O_2 \longrightarrow$

        fumarate + $2CO_2$ + $2H_2O$?

A. 4
B. 6
C. 7
D. 9
E. 10

28. Which of the following is included in oxidative phosphorylation?

A. Phosphoenolpyruvate + ADP $\longrightarrow$
                    pyruvate + ATP
B. glucose-6-phosphate + ADP $\longrightarrow$
                    glucose + ATP
C. ADP + phosphate $\longrightarrow$ ATP
D. UTP + ADP $\rightleftharpoons$ UDP + ATP
E. 2ADP $\rightleftharpoons$ ATP + AMP

29. A more negative oxidation-reduction system will reduce a more positive system relative to the standard hydrogen electrode. The most positive standard oxidation-reduction potential is shown by which of the following systems?

A. molecular oxygen system
B. NAD system
C. NADP system
D. flavoprotein system
E. cytochrome b system

30. Dinitrophenol, an uncoupler of oxidative phosphorylation,

A. inhibits NAD^+ - requiring reactions
B. inhibits cytochromes
C. inhibits respiration without affecting ATP synthesis
D. permits electron transport without ATP synthesis
E. inhibits ATP synthesis and respiration.

31. In mammals each of the following is a function of the tricarboxylic acid cycle EXCEPT:

A. net synthesis of oxaloacetate from acetyl CoA.
B. formation of α-ketoglutarate for amino acid biosynthesis.
C. generation of NADH and $FADH_2$.
D. metabolism of acetate to carbon dioxide and water.
E. oxidation of acetyl CoA produced primarily from glycolysis and oxidation of fatty acids.

32. Catabolism of fat produces more energy per gram than does carbohydrate or protein because:

A. fat yields much acetyl CoA that can enter the Krebs cycle
B. fat-metabolizing tissues produce more heat than carbohydrate-metabolizing tissues
C. fat has a higher (C + H)/O ratio than protein or carbohydrate
D. oxidative metabolism of fat goes most nearly to completion
E. the molecular weight of fatty acids is higher, on average, than that of amino acids or monosaccharides.

33. When electrons pass from succinate through $FADH_2$ and the electron transport system to oxygen, all of the following are true EXCEPT:

A. P:O ratio is 2
B. Coenzyme Q is involved
C. site I of H^+ pumping is bypassed
D. cytochrome c is involved
E. NADH dehydrogenase is reduced and re-oxidized.

34. What is the consequence of the oxidation of one mole of acetyl CoA via the TCA cycle and electron transport?

A. net consumption of 1 O_2
B. net production of 6 ATP
C. net production of 2 CO_2
D. net consumption of 1 oxaloacetate
E. net production of 2 GTP

35. Which of the following is able to phosphorylate ADP to ATP at the substrate level?

A. 3-phosphoglyceraldehyde
B. 3-phosphoglycerate
C. fructose-1,6-diphosphate
D. phosphoenolpyruvate
E. all of the above

36. Uncoupling of mitochondrial oxidative phosphorylation results in:

A. continued ATP formation, a halt in O_2 consumption
B. a slowing down of the Krebs cycle
C. inhibition of mitochondrial membrane ATPase
D. a halt in ATP formation but continued O_2 consumption
E. a halt in mitochondrial metabolism.

37. Which of the following reactions can phosphorylate ADP?

A. hexokinase
B. pyruvate kinase
C. phosphofructokinase
D. phosphorylase a
E. none of the above

38. Cytochrome oxidase reacts specifically with

A. carbon dioxide
B. periodic acid - Schiff reagent
C. parachloromercuribenzoate
D. cyanide
E. diisopropylfluorophosphate.

39. Which of the following can regulate the generation of ATP from glucose in muscle?

A. ATP
B. ADP
C. P_i
D. O_2
E. All of the above

40. How many ATP's are generated in the transfer of electrons from one NADH to oxygen?

A. 1
B. 2
C. 3
D. 4
E. 5

41. In addition to an enzyme complex, the conversion of pyruvate to acetyl CoA requires:

A. CoA, thiamin pyrophosphate and NAD^+
B. CoA, lipoic acid, thiamin pyrophosphate and FAD
C. CoA, lipoic acid, thiamin pyrophosphate, FAD and NAD^+
D. CoA, lipoic acid, biotin and ATP
E. CoA, ATP, NAD^+ and riboflavin.

42. Which one of the following enzymes contains lipoic acid bound in an amide linkage?

A. lactate dehydrogenase
B. phosphofructokinase
C. glycogen synthetase
D. ferrochelatase
E. pyruvate dehydrogenase

43. Which one of the following is required for the conversion of succinate to fumarate?

A. ATP
B. NAD^+
C. $NADP^+$
D. biotin
E. FAD

44 An enzyme

A. increases K_{eq} of the reaction
B. decreases K_{eq} of the reaction
C. increases free energy of reaction
D. decreases free energy of reaction
E. decreases activation energy of reaction.

45. In catalysis of a reaction, an enzyme changes the

A. rate only
B. rate and activation energy
C. activation energy only
D. K_{eq}
E. ΔG.

46. In an enzymatic reaction, $\Delta G^{o\prime}$ is

A. equal to the heat produced
B. equal to the activation energy
C. proportional to the concentration of enzyme
D. proportional to $-\log K_{eq}$
E. zero at standard conditions.

47. Cytochrome oxidase contains

A. cobalt
B. zinc
C. magnesium
D. vanadium
E. copper.

MATCHING: For each set of questions, choose the ONE BEST answer from the list of lettered options above it. An answer may be used one or more times, or not at all.

Questions 48 - 51: In the following, consider only changes at the substrate level; i.e., do not include the electron transport system.

A. Glucose → glycogen
B. Glycogen → fructose-6-phosphate
C. Glycogen → glucose-1-phosphate
D. Glucose-6-phosphate → glucose
E. Glucose → 3-phosphoglyceric acid

F. Pyruvate → lactate
G. Pyruvate → acetyl CoA
H. Citrate → α-ketoglutarate
I. α-Ketoglutarate → succinate
J. Succinate → oxaloacetate

48. Produces net quantity of high-energy phosphate.

49. Consumes net quantity of high-energy phosphate.

50. Both consumes and produces high-energy phosphate, in equal amounts, with net change = 0.

51. Does not occur in muscle.

Questions 52 - 54:

    A. Inner membrane of mitochondrion
    B. Outer membrane of mitochondrion
    C. Space between the two membranes
    D. Matrix
    E. Cytoplasm

52. Pyruvate dehydrogenase complex

53. Cytochrome respiratory chain

54. Glycolytic pathway

Questions 55 - 56:

    A. FAD
    B. lipoic acid
    C. NAD$^+$
    D. thiamin pyrophosphate
    E. all of the above

55. Forms an acyl thioester derivative during the enzymatic breakdown of pyruvate.

56. Cofactor for the pyruvate dehydrogenase complex.

VI. ANSWERS TO QUESTIONS ON ENERGETICS AND BIOLOGICAL OXIDATION

1. D	15. C	29. A	43. E
2. E	16. B	30. D	44. E
3. D	17. C	31. A	45. B
4. B	18. D	32. C	46. D
5. B	19. D	33. E	47. E
6. E	20. E	34. C	48. I
7. B	21. E	35. D	49. A
8. A	22. B	36. D	50. E
9. C	23. C	37. B	51. D
10. D	24. D	38. D	52. D
11. E	25. C	39. E	53. A
12. C	26. E	40. C	54. E
13. A	27. D	41. C	55. B
14. C	28. C	42. E	56. E

5. AMINO ACID METABOLISM

A. M. Chandler

I. FUNCTIONS OF AMINO ACIDS IN MAN

Amino acids are ingested in large amounts as structural components of dietary proteins. Unlike carbohydrate and fat, there are no large reserve stores of protein in the body. Thus, a continuous intake is required if tissue breakdown is to be avoided.

Amino acids are:
1. Precursors for the synthesis of proteins.
2. A source of energy under certain conditions.
3. Involved in the detoxification of drugs, chemicals and metabolic by-products.
4. Involved as direct neurotransmitters or as precursors to neurotransmitters.
5. Precursors to several peptide hormones and thyroid hormone.
6. Precursors to histamine, NAD and miscellaneous compounds of biological importance.

II. ESSENTIAL AND NON-ESSENTIAL AMINO ACIDS

All twenty amino acids are essential for life. A lack of a sufficient amount of any one of them leads to severe metabolic disruption and ultimate death.

Most microorganisms and plants are able to synthesize all 20 from glucose or CO_2 and NH_3. Mammals, however, including man, have during the process of evolution lost the ability to synthesize the carbon skeletons for several of the amino acids. Therefore, it is <u>essential</u> that these particular amino acids be obtained through the diet. Those amino acids that are not synthesized at a sufficient rate to meet demand are termed the **essential amino acids** and for man number ten. A useful mnemonic to assist in remembering the essential amino acids is **PVT TIM HALL:**

P- phenylalanine	T- tryptophan	H- histidine*
V- valine	I- isoleucine	A- arginine *
T- threonine	M- methionine	L- lysine
		L- leucine

Note that histidine and arginine are marked with asterisks. These amino acids are undoubtedly required for the infant and growing child, but it is less clear that they are essential for the normal, healthy adult.

III. NITROGEN BALANCE

The greatest portion of N intake is in the form of amino acids in the protein of the diet. After digestion, absorption and metabolic processing, the excess N derived from the NH_2 not required for growth or maintenance is excreted in the urine in the form of urea, NH_3 and other nitrogenous compounds. The normal, healthy adult is in "**nitrogen balance**" or "**equilibrium**". That is, the amount of N ingested in the diet over a given period of time equals that excreted in the urine and feces as excretory products.

Positive and Negative Nitrogen Balance. During pregnancy, infancy, childhood and in the recovery phase from a severe illness or surgery, the amount of N taken in and retained exceeds that excreted. The organism is said to be in a state of **positive nitrogen balance**. On the other hand, during starvation, <u>immediately</u> following severe trauma, surgery or other acute stress such as infections, N excretion exceeds intake and retention and the organism is in a state of **negative nitrogen balance**. A gradual, prolonged negative N balance is associated with **senescence**. Nitrogen balance is humorally controlled. Positive N balance is associated with growth hormone, insulin and with testosterone and other anabolic steroids. Negative N balance is associated with glucocorticoid action in mobilizing amino acids from muscle tissue.

IV. PROTEIN QUALITY

The dietary source of protein is also important for maintaining N balance. Not all proteins have the same biological value (BV). Proteins derived from animal sources have a high BV because they contain all the essential amino acids in the proper proportions. Plant proteins, on the other hand, usually are in lower tissue concentration and are harder to digest. In general, plant proteins are deficient in one or more essential amino acids, primarily lysine, tryptophan or methionine. In some third-world countries animal proteins are almost non-existent and protein intake may be limited to only one or two plant sources. The lack of a single essential amino acid leads to severe growth retardation in children and in adults to negative N balance. Growing children are particularly vulnerable and this is evidenced by the prevalence of **kwashiorkor** (See Chapter 10). Strict vegetarians can do well if they plan a diet containing a mixture of vegetable proteins, each one compensating for a defect in the other.

V. PROTEIN DIGESTION

A. Gastric Digestion

The first phase of protein digestion takes place in the stomach. **Gastrin**, a polypeptide hormone, is secreted into the blood by the antral gastric mucosa upon stimulation by foods. **Ethanol** is a particularly strong stimulator of gastrin release. Gastrin stimulates **chief cells** of the gastric mucosa to secrete the inactive proenzyme, **pepsinogen**, the **parietal cells** to secrete HCL and the **epithelial cells** to secrete **mucoproteins**. Once in contact with the very acidic environment of the stomach (pH< 5.0), a peptide fragment is cleaved from the pepsinogen molecule yielding the active protease, **pepsin**. Pepsin can then activate more pepsinogen autocatalytically. In addition to pepsinogen, other zymogens are secreted which yield pepsins B, C and D.

Pepsins (pH optima of 2.5) hydrolyze ingested proteins at sites involving aromatic amino acids, leucine and acidic amino acids. Because of the relatively short residence time of the stomach contents, digestion is limited. The partially digested, relatively large polypeptides then enter the duodenum of the small intestine.

B. Intestinal Digestion

Pancreatic Zymogens. The pancreas secretes several proenzymes into the duodenum along with a slightly alkaline fluid buffered to the pH optima of the active forms. The proenzymes include **trypsinogen, chymotrypsinogens, procarboxypeptidases** and **proelastase**.

Activations. Upon stimulation by the entrance of food into the intestine the intestinal mucosa secretes the enzyme **enterokinase** which acts on trypsinogen converting it to trypsin. Trypsin in turn activates more trypsinogen and the other proenzymes.

Trypsinogen $\xrightarrow{enterokinase}$ trypsin + peptide

Chymotrypsinogens $\xrightarrow{trypsin}$ chymotrypsins + peptides

Procarboxypeptidases $\xrightarrow{trypsin}$ carboxypeptidase + peptides

Proelastase $\xrightarrow{trypsin}$ elastase + peptide

Other Enzymes. The following brush border enzymes are secreted into the intestinal lumen but also work intracellularly:

Aminopeptidases: broad specificity; hydrolyze N-terminal amino acids.
Dipeptidases: hydrolyze dipeptides such as glycylglycine.
Prolinase: hydrolyzes peptides containing proline at the N-terminus.

VI. AMINO ACID ABSORPTION

Amino acid absorption is very rapid in the small intestine and is carried out by active transport mechanisms and is, therefore, an energy-expending process. Several specific transport mechanisms have been identified involving different classes of amino acids. These include those for:

1. small neutral amino acids
2. large neutral amino acids
3. basic amino acids
4. acidic amino acids
5. proline

Amino acids of the same class compete with one another for absorption sites.

The Gamma-Glutamyl Cycle (Figure 5-1). In addition to the transport mechanisms listed above, a general absorption mechanism involving all amino acids with a free amino group has been proposed, referred to as the **gamma-glutamyl cycle**. It provides a role for glutathione and explains the presence of 5-oxoproline in the urine. For every amino acid transported across the membrane by this proposed mechanism, three ATP's are consumed during the regeneration of glutathione.

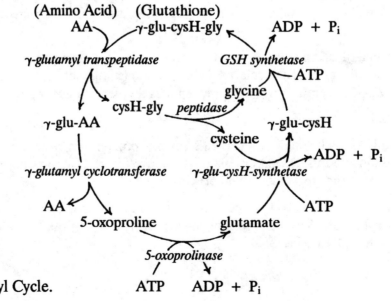

Figure 5-1.
The Gamma-Glutamyl Cycle.

In addition to amino acids, some small peptides are absorbed directly into the blood without hydrolysis. Very rarely, whole proteins are absorbed intact. The major route of entry into the blood is *via* the portal vein. Once in the blood amino acids are rapidly absorbed into cells. Liver and kidney take up the largest fraction. A blood-brain barrier exists for some amino acids, especially for glutamic acid.

VII. AMINO ACID DEGRADATION

In the breakdown of amino acids the first task the cell must accomplish is the removal of the alpha-amino groups. The two major mechanisms by which this is accomplished are **transamination** and **oxidative deamination**.

A. Transamination

Twelve amino acids can undergo transamination. These are ala, arg, asN, asp, cys, ile, leu, lys, phe, trp, tyr, and val. (Mnemonic: **VAL AT CAPITAL**).

The general reaction involved is:

$$\underset{\text{R-CH-COOH}}{\overset{\text{NH}_2}{|}} + \underset{\text{R}'\text{-C-COOH}}{\overset{\text{O}}{||}} \rightleftharpoons \underset{\text{R}'\text{-CH-COOH}}{\overset{\text{NH}_2}{|}} + \underset{\text{R-C-COOH}}{\overset{\text{O}}{||}}$$

The enzymes involved are called **transaminases** or **aminotransferases**. The reactions catalyzed by transaminases are freely reversible with equilibrium constants approaching 1.0. They can be found both in the mitochondria and the cytosol. The transfer of amino groups from most amino acids to α-ketoglutarate to form glutamate takes place in the cytosol. The glutamate formed can then enter the mitochondria via a special transport mechanism where it can undergo oxidative deamination or else form aspartic acid which can then reenter the cytoplasm.

Mechanism of transamination. All transaminases share a common reaction mechanism and use the same cofactor, **pyridoxal phosphate (PLP)**, which is derived from the vitamin, **pyridoxine (vitamin B$_6$)**. PLP is covalently bound to the enzyme via a Schiff's base linkage to an epsilon amino group of a specific lysine located in the active site. During transamination, a Schiff's base forms between the amino acids and PLP.

Details of the reaction.

First stage:

$$\underset{\underset{\text{AA}_1}{\text{R}_1\text{-CH-COOH}}}{\overset{\text{NH}_2}{|}} + \underset{\text{PLP-E}}{\underset{\text{E}}{\overset{\text{HC=O}}{|}}} \underset{\longleftarrow}{\overset{\text{H}_2\text{O}}{\nearrow}} \underset{\underset{\text{Aldimine}}{\text{R}_1\text{-CH-COOH}}}{\overset{\text{N=CH-E}}{|}} \longleftrightarrow \underset{\underset{\text{Ketimine}}{\text{R}_1\text{-C-COOH}}}{\overset{\text{N-CH}_2\text{-E}}{||}} \underset{\longleftarrow}{\overset{\text{H}_2\text{O}}{\nwarrow}} \underset{\text{PMP-E}}{\overset{\text{CH}_2\text{-NH}_2}{|}} \underset{\text{E}}{} + \underset{\underset{\alpha\text{-Ketoacid}_1}{\text{R}_1\text{-C-COOH}}}{\overset{\text{O}}{||}}$$

Second stage:

$$\underset{\underset{\alpha\text{-Ketoacid}_2}{\text{R}_2\text{-C-COOH}}}{\overset{\text{O}}{||}} + \underset{\text{PMP-E}}{\overset{\text{CH}_2\text{-NH}_2}{|}} \underset{\longleftarrow}{\overset{\text{H}_2\text{O}}{\nearrow}} \underset{\underset{\text{Ketimine}}{\text{R}_2\text{-C-COOH}}}{\overset{\text{N-CH}_2\text{-E}}{||}} \longleftrightarrow \underset{\underset{\text{Aldimine}}{\text{R}_2\text{-CH-COOH}}}{\overset{\text{N=CH-E}}{|}} \underset{\longleftarrow}{\overset{\text{H}_2\text{O}}{\nwarrow}} \underset{\text{PLP-E}}{\overset{\text{HC=O}}{|}} \underset{\text{E}}{} + \underset{\underset{\text{AA}_2}{\text{R}_2\text{-CH-COOH}}}{\overset{\text{NH}_2}{|}}$$

B. Oxidative Deamination

L-glutamate is the ultimate product from the majority of the transaminations that occur. Glutamate then enters mitochondria where the following reaction occurs:

$$\text{L-glutamate} + NAD^+ + H_2O \xrightarrow{\textit{glutamate dehydrogenase}} \alpha\text{-ketoglutarate} + NH_4^+ + NADH$$

The reaction is catalyzed by *glutamate dehydrogenase* found in both the mitochondria and cytoplasm. The enzyme can use both NAD^+ and $NADP^+$ but NAD^+ is preferred in the catabolic direction. The NADH formed within the mitochondrion can enter the electron transport system and yield 3 ATP.

Glutamate dehydrogenase is a very complex enzyme made up of six identical 56,000-dalton subunits. It is regulated allosterically by a number of effectors.

Inhibitors: ATP, GTP, NADH
Activators: ADP, GDP, specific amino acids
Hormones: thyroxine and steroid hormones

C. Amino Acid Oxidases

L-amino acid oxidase is found in liver, kidney and snake venom. It is involved primarily in the deamination of lysine (E = oxidase molecule):

$$\text{L-amino acid} + H_2O + \text{E-FMN} \longrightarrow \alpha\text{-ketoacid} + NH_3 + \text{E-FMNH}_2$$

D-amino acid oxidase:

$$\text{D-amino acid} + H_2O + \text{E-FAD} \longrightarrow \alpha\text{-ketoacid} + NH_3 + \text{E-FADH}_2$$

The flavin nucleotides (FMN and FAD) are derived from **riboflavin (vitamin B$_2$)**. The regeneration of the flavin nucleotides is carried out utilizing molecular oxygen yielding H_2O_2 which is rapidly degraded by *catalase*.

VIII. THE UREA CYCLE (KREBS-HENSELEIT CYCLE)

Ammonia is considered to be toxic to cells of the CNS in higher organisms and must be kept at low concentrations. In many aquatic animals like fish and tadpoles, NH_3 diffuses directly into the surrounding water through gills. These animals are called **ammonotelic**. Others such as birds and reptiles excrete the ammonia as uric acid (**uricotelic**). Man and most other terrestrial vertebrates excrete NH_3 as urea (**ureotelic**).

Steps in the production of urea are shown in Figure 5-2, and described in sections A - E (below and next page):

A. *Carbamoyl Phosphate Synthetase.*

The enzyme is mitochondrial, and responds to allosteric activation by N-acetylglutamate.

$$2\,ATP + CO_2 + NH_3 + H_2O \longrightarrow \overset{\displaystyle O}{\overset{\displaystyle \|}{H_2N\text{-}C\text{-}O\text{-}PO_3}} + 2\,ADP + P_i$$

Carbamoyl
phosphate

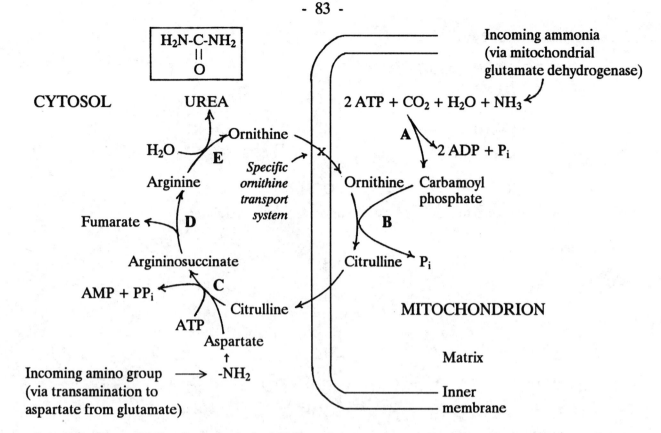

Figure 5-2. The Urea Cycle. Letters in boldface refer to enzymes described in sections A - E in text.

B. *Ornithine Carbamoyl Transferase*

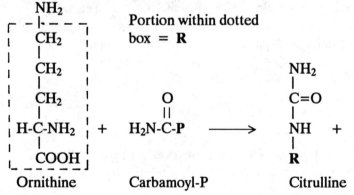

C. *Argininosuccinate Synthetase*

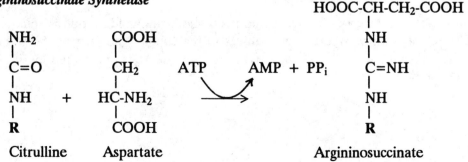

D. *Argininosuccinate Lyase*

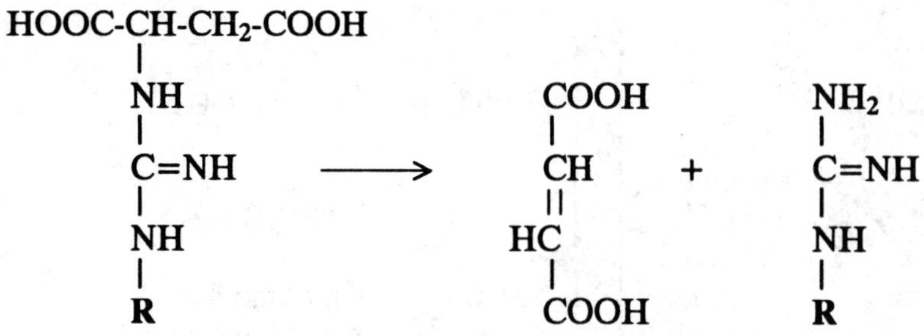

$$
\begin{array}{ccc}
\text{HOOC-CH-CH}_2\text{-COOH} & & \\
| & & \\
\text{NH} & \text{COOH} & \text{NH}_2 \\
| & | & | \\
\text{C=NH} \longrightarrow & \text{CH} + & \text{C=NH} \\
| & || & | \\
\text{NH} & \text{HC} & \text{NH} \\
| & | & | \\
\text{R} & \text{COOH} & \text{R} \\
\text{Argininosuccinate} & \text{Fumarate} & \text{Arginine}
\end{array}
$$

E. *Arginase*

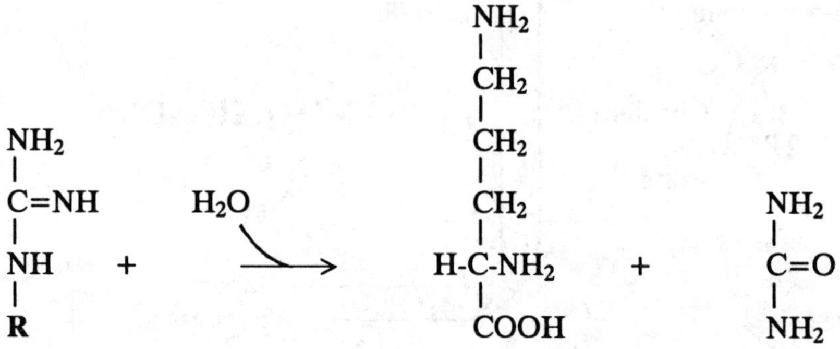

$$
\begin{array}{cccc}
 & & \text{NH}_2 & \\
 & & | & \\
 & & \text{CH}_2 & \\
 & & | & \\
 & & \text{CH}_2 & \\
\text{NH}_2 & & | & \\
| & & \text{CH}_2 & \text{NH}_2 \\
\text{C=NH} \quad \text{H}_2\text{O} & & | & | \\
| \quad\quad\quad & \longrightarrow & \text{H-C-NH}_2 + & \text{C=O} \\
\text{NH} \quad + & & | & | \\
| & & \text{COOH} & \text{NH}_2 \\
\text{R} & & & \\
\text{Arginine} & & \text{Ornithine} & \text{Urea}
\end{array}
$$

Overall Reaction:

$$2\,NH_3 + CO_2 + 3\,ATP + 3\,H_2O \longrightarrow Urea + 2\,ADP + AMP + 2\,P_i + PP_i$$

The urea cycle is intimately tied to the citric acid cycle by way of fumarate and aspartate. One N is derived from NH_3 and one from aspartate. Each of the enzymes of the urea cycle have been identified with specific genetic defects. Upon ingestion of proteins such patients show hyperammonemia, lethargy, vomiting and other signs of CNS disturbance.

IX. EXCRETION OF FREE AMMONIA

In addition to excretion as urea, **NH_3** can be removed from cells by transferring it to the gamma carboxyl of glutamate to form **glutamine**. Glutamine is non-toxic and can pass through the blood-brain barrier.

$$\text{Glutamate} + ATP + NH_3 \xrightarrow{\substack{glutamine \\ synthetase}} \text{glutamine} + ADP + P_i$$

Glutamine is transported in the blood either to the liver where the amide group is removed and converted to urea, or else to the kidneys where it encounters in the tubules the enzyme, *glutaminase*. Glutaminase hydrolyzes the glutamine to glutamate and NH_3. The glutamate is reabsorbed by the tubules and the ammonia is excreted in the urine.

X. DEGRADATION OF THE CARBON SKELETONS

After removal of the α-amino groups, the carbon skeletons of the 20 amino acids undergo a series of reactions that result in products that are members of the glycolytic pathway, the citric acid cycle or ketone bodies. There are only seven of these: pyruvate, acetyl CoA, acetoacetyl CoA, α-ketoglutarate, succinyl CoA, fumarate and oxaloacetate.

Table 5-1 outlines the catabolism of the 20 amino acids.

Amino Acid	Products	Number of Enzymatic Steps	Cofactors	Glycogenic or Lipogenic
Alanine	Pyruvate	1	PLP	G
Glycine	Pyruvate	2	N^5,N^{10}-methylene THFA	G
Serine	Pyruvate	1		G
Cysteine	Pyruvate	2	PLP, NADH	G
Threonine	Pyruvate	3		G
Aspartic Acid	Oxaloacetate	1	PLP	G
Asparagine	Oxaloacetate	2	PLP	G
Histidine	α-ketoglutarate	5	THFA, PLP	G
Glutamic acid	α-ketoglutarate	1	PLP	G
Glutamine	α-ketoglutarate	2	PLP	G
Arginine	α-ketoglutarate	4	PLP, NAD	G
Proline	α-ketoglutarate	4	O_2, PLP	G
Methionine	succinyl CoA	9	ATP, CoA, NAD, biotin, Vit B_{12}	G
Valine	succinyl CoA	10	PLP, NAD, CoA, Vit B_{12}	G
Isoleucine	succinyl CoA	9	PLP, NAD, CoA, FAD, biotin, Vit B_{12}	G,L
Leucine	acetyl CoA, acetoacetyl CoA	6	thiamin PP, lipoic acid, biotin, Vit B_{12}, PLP, CoA, NAD, FAD	L
Phenylalanine	acetoacetyl CoA, fumarate	7	O_2, NADPH, tetrahydrobiopterin	G,L
Tyrosine	acetoacetyl CoA, fumarate	6	O_2, NADPH, tetrahydrobiopterin	G,L
Tryptophan	acetoacetyl CoA, alanine	9	O_2, NADPH, NAD	G,L
Lysine	acetoacetyl CoA	9	NADPH, NAD, NADP, PLP, CoA, FAD	G,L

Table 5-1. Catabolism of the 20 Amino Acids

XI. ONE-CARBON FRAGMENT METABOLISM

In the degradation and synthesis of amino acids (and many other compounds), there is often a need to remove, add or rearrange 1-C units. With the exception of the decarboxylases which remove 1-C units as CO_2 or HCO_3^-, all 1-carbon transfers require the participation of cofactors. The precursors of these cofactors must be supplied by the diet, either as vitamins or as essential amino acids.

A. Tetrahydrofolate (THFA) (Tetrahydropteroylglutamate)

Structure. This compound is the most versatile of the carriers of 1-C fragments and carries them at several levels of oxidation, from the most reduced, methyl, to the most oxidized, methenyl:

$-CH_3$	methyl
$-CH_2-$	methylene
$-CH=O$	formyl
$-CH=NH$	formimino
$-CH=$	methenyl

Tetrahydrofolate is composed of three units: (a) a pteridine derivative, (b) p-aminobenzoic acid (PABA) and (c) glutamic acid. The 1-C fragments are carried either on N^5 or N^{10} or as a bridge between both (Figure 5-3). The **sulfa drugs** act by inhibiting the bacterial enzymes responsible for coupling the PABA moiety to the other components to form the THFA molecule. The bacteria are than inhibited in their ability to carry out 1-C transfers. The sulfa drugs are structural analogues of the PABA moiety. Other drugs such as **methotrexate** also act through interference with the action of tetrahydrofolate. (See Chapter 11)

Metabolism. The several forms of 1-C fragments carried by THFA are interconvertible one to the other (Figure 5-3). Each reaction is reversible, therefore, 1-C fragments can be donated or accepted by each species, creating an equilibrium mixture of 1-C fragments in different oxidation states.

THFA is involved in transferring a methyl group to Vitamin B_{12} in the reactions converting homocysteine to methionine. This is the only methyl transfer involving THFA. All other methylations involve S-adenosylmethionine.

B. S-Adenosylmethionine (SAM)

SAM is the cofactor involved in transmethylations. The general reaction is:

$$SAM + R(acc) \xrightarrow[\text{transferase}]{\text{methyl}} RCH_3 + \text{S-adenosylhomocysteine}$$

The structure of SAM and its formation and metabolism are shown in Figure 5-4.

C. Biotin

All carboxylations via transcarboxylases use **biotin** as a cofactor. Each carboxylation reaction involves a specific transcarboxylase.

$$E\text{-biotin} + ATP + HCO_3^- \longrightarrow E\text{-biotin-}CO_2 + ADP + P_i$$
$$E = \text{transcarboxylase}$$

The biotin is attached to the transcarboxylase through the ϵ-amino group of a lysine in the active center. When "activated" the CO_2 is attached to a nitrogen of the biotin molecule.

Biotin is a vitamin. It is bound tightly and irreversibly by **avidin,** a protein found in significant quantities in raw egg white.

- 87 -

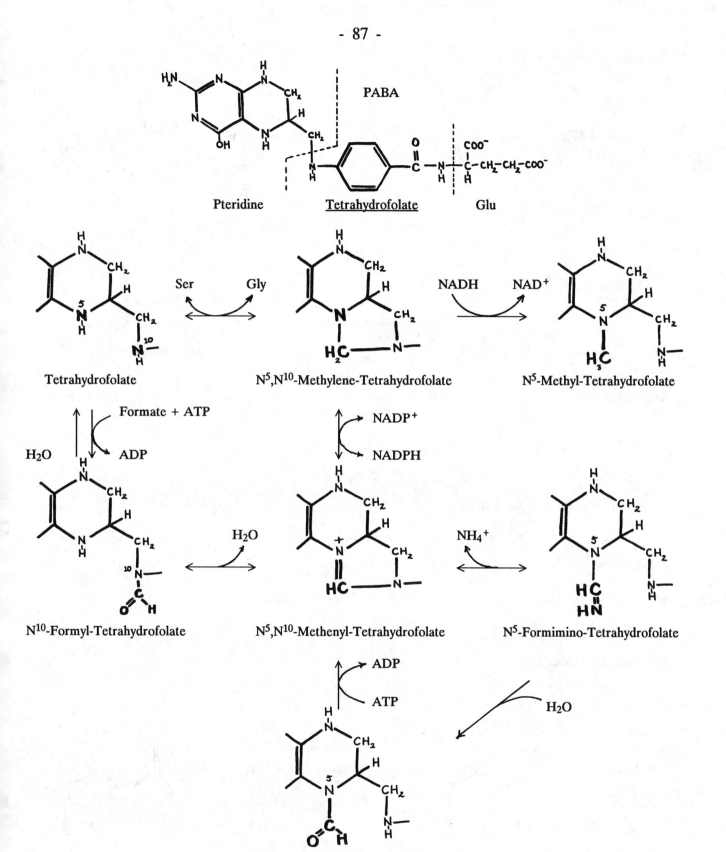

Figure 5-3. Structure of Tetrahydrofolic Acid and the Metabolism of the Various C-1-Carrying Forms.

1. Synthesis of SAM from Methionine:

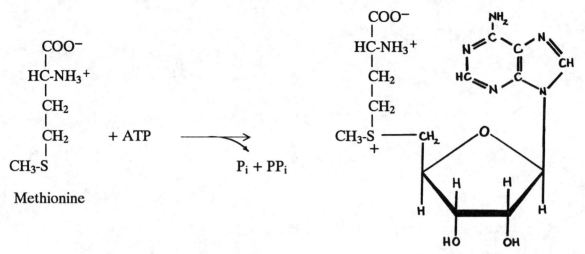

2. Transfer of Methyl Group:

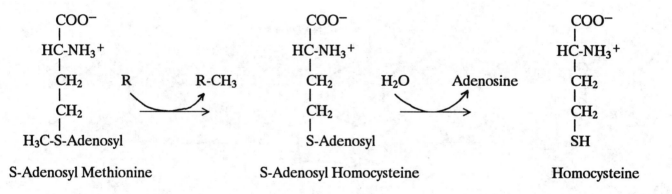

3. Re-synthesis of Methionine:

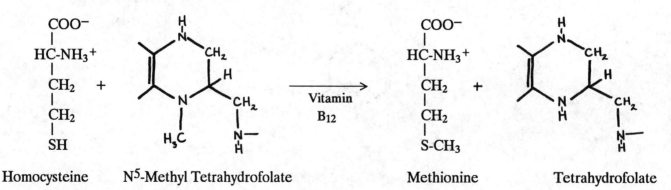

Figure 5-4. S-Adenosyl Methionine: Structure, Formation and Metabolism.

D. Cyanocobalamin (Vitamin B$_{12}$)

Vitamin B$_{12}$ (cyanocobalamin) has a complex ring structure and has as an essential component an atom of the trace metal, **cobalt**. It participates in the transfer of methyl groups from N^5-methyltetrahydrofolate to homocysteine to regenerate methionine. It is also involved in certain one-carbon rearrangement reactions as in the conversion of methylmalonyl CoA to succinyl CoA in the latter stages of the degradation of Met, Val and Ile. An inability to obtain enough Vitamin B$_{12}$ in the diet or an inability to absorb it in the intestinal tract leads to the condition of **pernicious anemia**. The appearance of methylmalonic acid in the blood and urine can also be caused by vitamin B$_{12}$ deficiency.

XII. METABOLISM OF PHENYLALANINE AND TYROSINE

Several genetic defects exist involving the enzymes of phenylalanine and tyrosine degradation. Tyrosine is also the precursor to several neurohormones and to melanin. Therefore, the metabolism of these amino acids will be covered in slightly more detail.

A. Degradation of Phenylalanine and Tyrosine

The scheme for the degradation of phenylalanine and tyrosine is shown in Figure 5-5.

Phenylketonuria. The most common genetic disturbance in the metabolism of these amino acids occurs at the first step, the oxidation of phenylalanine to tyrosine by *phenylalanine hydroxylase (phenylalanine 4-monooxygenase)*. The absence of this enzyme or a defect in its functioning caused by defects in other components of the system leads to the accumulation of **phenylpyruvate** causing the condition known as **phenylketonuria**. This condition has a frequency of about 1/15,000 and if untreated is characterized by mental retardation, CNS damage and hypopigmentation. If diagnosed early, some affected children may be spared from the major damaging effects of phenylpyruvate accumulation by placing them on a diet low in phenylalanine, substituting the phenylalanine with tyrosine.

An essential cofactor for phenylalanine hydroxylase is **tetrahydrobiopterin** which is also oxidized, during the oxidation of phenylalanine to tyrosine, to **dihydrobiopterin**. The dihydro form must be reduced back to the tetrahydro form to sustain the overall conversion reaction. The enzyme carrying out this reduction uses NADPH as a cofactor. Some cases of phenylketonuria are the result of defects in this enzymatic step or in steps related to the synthesis of tetrahydrobiopterin.

Alcaptonuria. A second defect in tyrosine catabolism has been observed in persons with the condition **alcaptonuria**. A diminished activity of the enzyme converting **homogentisic acid** to 4-maleylacetoacetate leads to the accumulation of homogentisic acid in the urine which is spontaneously oxidized by atmospheric oxygen to produce a black or dark-colored urine. This condition is relatively benign.

B. Conversion of Tyrosine to Neurohormones and Melanin

Figure 5-6 shows the conversion of tyrosine to DOPA, dopamine, norepinephrine, epinephrine and melanin.

DOPA (dihydroxyphenylalanine) is a key intermediate in these pathways. Genetic defects in the conversion of DOPA to **melanin** causes **albinism**, a condition of very severe hypopigmentation. **Dopamine** and **norepinephrine** are neurotransmitters. **Epinephrine** is synthesized primarily in the adrenal medulla. Disturbances in the functioning of CNS pathways using dopamine as a transmitter have been linked to the condition of **schizophrenia**. A lack of sufficient dopamine production in certain brain structures like the *substantia nigra* or the destruction of this structure by toxic compounds leads to **Parkinson's Disease** or the

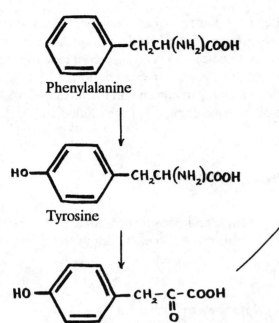
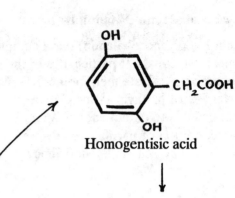

Phenylalanine

Tyrosine

4-Hydroxyphenylpyruvic acid

Homogentisic acid

4-Maleylacetacetic acid

4-Fumarylacetoacetic acid

Acetoacetic acid   Fumaric acid

Acetyl CoA

Figure 5-5. Degradation of Phenylalanine and Tyrosine.

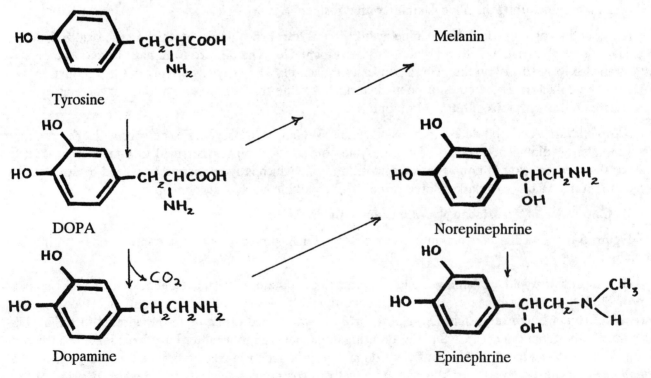

Tyrosine

DOPA

Dopamine

Melanin

Norepinephrine

Epinephrine

Figure 5-6. Metabolism of Tyrosine.

Parkinsonian syndrome. Administration of large quantities of L-DOPA can reverse many of the symptoms in some cases.

Norepinephrine is found primarily at nerve endings of the adrenergic (sympathetic) nervous system.

XIII. GENERAL PRECURSOR FUNCTIONS OF AMINO ACIDS

In addition to serving as precursors to proteins, amino acids also act as precursors to many other compounds of biological importance. Table 5-2 lists a few of these compounds and the amino acid(s) from which they are derived.

Compound	Amino Acid Precursor(s)
Neurotransmitters	
many amino acids serve directly as neurotransmitters	
dopamine, epinephrine,norepinephrine	Phe, Try
serotonin (5-hydroxytryptamine)	Trp
GABA (γ-aminobutyric acid)	Glu
acetylcholine (via ethanolamine)	Ser
Miscellaneous compounds	
indole acetic acid (plant hormone)	Trp
creatine and phosphocreatine	Met, Gly, Arg
spermine, spermidine	Met, Arg (Orn)
histamine	His
thyroxine	Tyr
heme	Gly
polypeptide hormones	all 20 amino acids
NAD	Trp
taurine	Cys
carnitine	Lys
purine bases	Gly
carnosine, anserine	His
Coenzyme A	Cys

Table 5-2. Important Derivatives of Amino Acids

XIV. REVIEW QUESTIONS ON AMINO ACID METABOLISM

> **DIRECTIONS:** For each of the following multiple-choice questions (1 - 52), choose the ONE BEST answer.

1. A man is in negative nitrogen balance when his:

A. dietary nitrogen drops below the recommended daily allowance.
B. fecal nitrogen excretion exceeds his urinary nitrogen excretion.
C. diet contains more nonessential amino acids than essential amino acids.
D. urinary nitrogen excretion exceeds his dietary nitrogen intake.
E. dietary nitrogen intake exceeds his urinary nitrogen excretion.

2. The intravenous injection of a labeled, radioactive amino acid to a mature animal will lead to:

A. no incorporation of labeled amino acid into protein
B. incorporation of equivalent amounts of labeled amino acid into all the proteins of the body
C. incorporation of different amounts of labeled amino acid into various proteins of the body
D. rapid and complete excretion of the labeled amino acid
E. excretion of an equal amount of unlabeled amino acid

3. If a mature animal in nitrogen balance is placed on a diet deficient only in phenylalanine, which of the following conditions is most likely to occur?

A. Nitrogen balance will become negative and remain that way so long as the deficiency exists.
B. Nitrogen balance will become negative temporarily, but the individual will adapt and nitrogen balance will gradually return to zero.
C. Nitrogen intake will continue to equal nitrogen excretion (balance= 0).
D. Nitrogen balance will become positive and remain that way so long as the deficiency exists.
E. Nitrogen balance will become positive temporarily, but the individual will adapt and nitrogen balance will gradually return to zero.

4. Which of the following enzyme pairs is involved in the conversion of amino acid nitrogen into two compounds that directly provide the urea nitrogen?

A. glutamic-oxaloacetic transaminase and diamine oxidase
B. L-amino acid oxidase and racemase
C. serine dehydratase and glutamate dehydrogenase
D. carbamoyl phosphate synthetase and glutamic-oxaloacetic transaminase
E. glutamine synthetase and glutaminase

5. In the regeneration of methionine from homocysteine, which of the following cofactors is/are involved?

A. lipoic acid
B. retinoic acid
C. biotin and thiamin pyrophosphate
D. cofactors derived from tetrahydrofolic acid and vitamin B_{12}
E. cofactors derived from vitamins E and K

6. The rate-limiting reaction in the urea cycle is that catalyzed by:

A. argininosuccinase
B. argininosuccinate synthetase
C. arginase
D. ornithine transcarbamoylase
E. carbamoyl phosphate synthetase

7. The utilization of ammonia for synthesis of the α-amino group of non-essential amino acids:

A. is dependent on the action of glutamate dehydrogenase
B. is achieved by reversal of the urea cycle
C. is mediated by carbamoyl phosphate
D. is dependent on intestinal bacteria
E. requires the participation of glutamine

8. Which one of the following is an essential amino acid?

A. asparagine
B. glycine
C. glutamic acid
D. methionine
E. serine

9. The combination of which of the following enzymatic activities provides the major route of flow of nitrogen from amino acids to ammonia in man?

A. amino acid oxidases and glutamate dehydrogenase
B. glutamate dehydrogenase and glutaminase
C. transaminases and glutamate dehydrogenase
D. transaminases and glutaminase
E. glutaminase and amino acid oxidases

10. An amino acid which is both ketogenic and glucogenic is:

A. tyrosine
B. alanine
C. leucine
D. glutamic acid
E. histidine

11. Transport of amino acids by the gamma-glutamyl cycle involves the direct participation of which one of the following components?

A. GTP
B. aspartic acid
C. glutathione
D. glutamine
E. glucose

12. Phenylketonuria is a genetic defect caused by the absence of the enzyme:

A. α-keto acid decarboxylase
B. tyrosinase
C. homogentisic acid oxidase
D. phenylalanine hydroxylase
E. alanine transaminase

13. Dopamine is a neurohormone in brain. Its precursor is:

A. cysteine
B. phenylalanine
C. asparagine
D. tryptophan
E. lysine

14. The enzyme enterokinase is important in the intestinal digestion of dietary protein because it converts:

A. pepsinogen to pepsin.
B. procarboxypeptidase A to carboxypeptidase A.
C. trypsinogen to trypsin.
D. procarboxypeptidase B to carboxypeptidase B.
E. chymotrypsinogen to chymotrypsin.

15. Excess lysine added to a diet might cause decreased absorption of which of the following amino acids?

A. Proline
B. Arginine, histidine
C. Phenylalanine, tyrosine
D. Aspartic acid, glutamic acid
E. Glycine, alanine

16. A diet that consists exclusively of corn and no other source of protein, will lead to a state of:

A. nitrogen equilibrium
B. positive nitrogen balance
C. negative nitrogen balance
D. grace
E. confusion

17. Which one of the following intermediates directly links the urea cycle with the TCA cycle?

A. Acetyl-CoA
B. Arginine
C. Fumarate
D. Glutamate
E. Pyruvate

18. Which level of oxidation of one carbon units CANNOT be accommodated by the folic acid coenzymes?

A. CO_2
B. Formate
C. $-CH_2-OH$
D. $-CH_3$
E. $-HC=0$

19. Which one of the following groups consists entirely of amino acids which are nutritionally essential for humans?

A. Valine, isoleucine, tyrosine, arginine.
B. Leucine, methionine, isoleucine, alanine.
C. Glutamic acid, arginine, cysteine, tryptophan.
D. Valine, isoleucine, tyrosine, lysine.
E. Lysine, tryptophan, phenylalanine, threonine.

20. The coenzyme involved in transaminations and many other amino acid transformations is derived from:

A. niacin
B. pyridoxine
C. flavins
D. thiamine
E. vitamin B_{12}

21. Compounds important to neural function resulting, in full or part, from amino acid decarboxylations include:

A. norepinephrine
B. gamma-amino butyrate
C. serotonin
D. acetyl choline
E. all of the above

22. A reaction type in which S-adenosylmethionine participates is:

A. transmethylation.
B. transformylation.
C. transadenylation.
D. transamination.
E. transcription.

23. A departure from nitrogen balance is observed in which of the following conditions?

A. obesity
B. pregnancy
C. menstruation
D. moderate exercise
E. sleep

24. The major nitrogen-containing compound(s) excreted in the urine is:

A. amino acids
B. uric acid
C. creatinine
D. urea
E. NH_4^+

25. Glutamic acid dehydrogenase carries out the oxidative deamination of glutamate in the mitochondria. The products of this reaction when the cell is in a low energy state are:

A. glutamate, NH_4^+, NADH
B. α-ketoglutarate, NH_4^+, NAD^+
C. α-ketoglutarate, NH_4^+, NADH
D. α-ketoglutarate, NH_4^+, ATP
E. α-ketoglutarate, NH_4^+, $NADP^+$

26. After glucose labeled in the C-6 position with ^{14}C was administered to a mouse, the animal's proteins became labeled. Which of the following amino acids, derived from these proteins, would NOT be labeled?

A. proline
B. leucine
C. aspartate
D. cystine
E. alanine

27. The sources of the nitrogen atoms in urea come from which of the following urea cycle compounds?

A. carbamoyl phosphate only
B. carbamoyl phosphate and glutamine
C. aspartic acid and glutamine
D. aspartic acid and carbamoyl phosphate
E. lysine

28. The major source of urinary ammonia produced by the kidney is:

A. leucine
B. glycine
C. glutamine
D. asparagine
E. glutamic acid

29. Upon degradation, serine, alanine and cysteine are likely to be converted to:

A. α-ketoglutarate
B. pyruvate
C. fumarate
D. succinate
E. phenylpyruvate

30. Pathways for the synthesis of pyrimidines, urea, and citrulline have in common a requirement for:

A. acetate
B. carbamoyl phosphate
C. tetrahydrofolate
D. propionate
E. pyruvate

31. In maple syrup urine disease (now called branched chain aminoaciduria), the defective metabolic step involves:

A. an oxidative decarboxylation
B. an amino acid transaminase
C. a methionine deficiency in the diet
D. an amino acid hydroxylase
E. amino group fixation to carbon skeletons

32. The major non-protein nitrogenous component of blood is:

A. urea
B. ammonia
C. a purine
D. an amino acid
E. uric acid

33. The major source of ammonia (NH_4^+) in fresh urine is from:

A. the hydrolysis of glutamine by glutaminase
B. the hydrolysis of urea by urease.
C. the oxidation of amino acids by L-amino acid oxidase
D. the oxidation of amines by an amine oxidase
E. deamination of aspartate by aspartate ammonia lyase

34. The phosphorylated form of creatine is:

A. a source of high energy phosphate for ATP formation in muscle
B. a component of the urea cycle
C. excreted by the kidney
D. an important hormone
E. none of the above

35. Which of the following is not involved in the biosynthesis of creatine phosphate?

A. pyridoxal phosphate
B. ATP
C. methionine
D. glycine
E. arginine

36. Dietary methionine can be replaced <u>entirely</u> by which one of the following compounds?

A. threonine
B. homocysteine
C. folic acid
D. cysteine
E. creatine

37. Which one of the following amino acids is non-essential as a human nutrient?

A. lysine
B. phenylalanine
C. valine
D. threonine
E. proline

38. Which one of the following must be supplied to the normal human adult in order to maintain nitrogen balance?

A. threonine
B. alanine
C. lysine
D. tyrosine
E. serine

39. Which of the following enzymes is most important in the process of dietary protein digestion (and especially zymogen activation)?

A. enterokinase
B. prolidase
C. carboxypeptidase B
D. pepsin
E. chymotrypsin

40. In the genetic defect, homocystinuria, the defective enzyme is thought to be:

A. methionine demethylase
B. cystathionine synthetase
C. cystathionase
D. S-adenosylhomocysteine hydrolase
E. a kidney enzyme involved in amino acid transport

41. All of the following occur in humans EXCEPT:

A. serine → cysteine
B. phenylalanine → tyrosine
C. glutamate → proline
D. oxaloacetate → lysine
E. homocysteine → methionine

42. Tryptophan is <u>not</u> used in the biosynthesis of:

A. niacin
B. serotonin
C. norepinephrine
D. melatonin
E. indoles

43. Biosynthesis of glucose from aspartate involves:

A. dephosphorylation
B. transamination
C. aldol condensation
D. carboxylation
E. all of the above

44. A direct donor of a nitrogen atom to urea is:

A. ornithine
B. methionine
C. aspartic acid
D. glutamic acid
E. creatinine

45. A <u>positive</u> nitrogen balance is most likely in:

A. a growing child
B. a healthy adult
C. a senescent adult
D. a child on a lysine-deficient diet
E. an adult on a phenylalanine deficient diet

46. The primary site of urea synthesis is in the:

A. kidney
B. skeletal muscles
C. liver
D. small intestine
E. brain

47. Which amino acid is directly involved in transfer of 1-carbon fragments?

A. methionine
B. tryptophan
C. proline
D. tyrosine
E. threonine

48. Which enzymatic step is <u>not</u> involved in the biosynthesis of epinephrine?

A. an amino oxidation
B. a methylation
C. an aliphatic hydroxylation
D. an aromatic hydroxylation
E. a decarboxylation

49. Conversion of ornithine to citrulline is a step in the synthesis of:

A. arginine
B. cysteine from methionine
C. tyrosine from glucose
D. urea from NH_3
E. A and D are correct

50. Which of the following would most strongly initiate gastrin release upon ingestion?

A. ethanol
B. fats
C. glucose
D. protein
E. DNA

51. Zymogen molecules are converted to active enzymes as a result of:

A. enzymatic phosphorylation
B. activation by ATP
C. enzyme adenylate formation
D. limited proteolysis
E. substrate binding

52. Hormones derived from amino acid precursors include those of the:

A. pancreas
B. thyroid gland
C. adrenal medulla
D. parathyroid glands
E. all of the above

MATCHING: For each set of questions, choose the ONE BEST answer from the list of lettered options after it. An answer may be used one or more times, or not at all.

Questions 53 - 56: Aromatic amino acid metabolism involves the following pathway:

phenylalanine
A ↓
     B
tyrosine ⟶ melanin
C ↓
p-hydroxyphenylpyruvate
D ↓
homogentisic acid
E ↓
maleoyl acetoacetic acid

Match the condition below with an enzymatic block (lettered in diagram at left).

53. Tyrosinosis.

54. Albinism.

55. Phenylketonuria.

56. Alkaptonuria.

Questions 57 - 64: Which of the following is most directly involved as a precursor in the biosynthesis of the metabolites in questions 57 - 64?

    A. tyrosine
    B. tryptophan
    C. serine
    D. glutamic acid
    E. glycine

57. Porphyrins.

58. γ-Aminobutyric acid.

59. Serotonin.

60. Epinephrine.

61. Acetylcholine.

62. Thyroxine.

63. Niacin.

64. Creatine.

Questions 65 - 67:

    A. α-ketoglutaric acid
    B. glutamic acid
    C. aspartic acid
    D. acetyl CoA
    E. pyruvic acid

65. Can be deaminated enzymatically to form oxaloacetc acid.

66. Is formed directly from alanine by transamination.

67. Deamination directly yields α-ketoglutarate.

Questions 68 - 71:

    A. enterokinase
    B. chymotrypsinogen
    C. carboxypeptidase
    D. pepsin
    E. trypsin

68. An intestinal enzyme which initiates zymogen activation.

69. An exopeptidase.

70. A protease with a pH optimum less than 3.0.

71. A precursor to an endopeptidase.

Questions 72 - 76:

    A. pyridoxal phosphate
    B. tetrahydrofolate
    C. adenosine 5'-triphosphate
    D. a flavin nucleotide
    E. NADH

72. The interconversion of serine and glycine.

73. Transaminations.

74. The formation of S-adenosyl methionine.

75. The conversion of δ-pyrroline-5-carboxylate to proline.

76. The oxidation of amines.

XV. ANSWERS TO QUESTIONS ON AMINO ACID METABOLISM

1.	D	20.	B	39.	A	58.	D
2.	C	21.	E	40.	B	59.	B
3.	A	22.	A	41.	D	60.	A
4.	D	23.	B	42.	C	61.	C
5.	D	24.	D	43.	E	62.	A
6.	E	25.	C	44.	C	63.	B
7.	A	26.	B	45.	A	64.	E
8.	D	27.	D	46.	C	65.	C
9.	C	28.	C	47.	A	66.	E
10.	A	29.	B	48.	A	67.	B
11.	C	30.	B	49.	E	68.	A
12.	D	31.	A	50.	A	69.	C
13.	B	32.	A	51.	D	70.	D
14.	C	33.	A	52.	E	71.	B
15.	B	34.	A	53.	C	72.	B
16.	C	35.	A	54.	B	73.	A
17.	C	36.	B	55.	A	74.	C
18.	A	37.	E	56.	E	75.	E
19.	E	38.	C	57.	E	76.	D

6. PORPHYRINS

Thomas Briggs

I. STRUCTURE AND CHEMISTRY

Porphyrins: cyclic tetrapyrroles found in **all aerobic cells**. With a system of conjugated double bonds, they absorb light strongly and are intensely colored, as in blood.

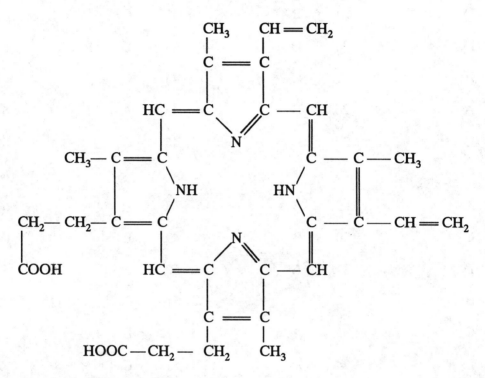

Figure 6-1. Protoporphyrin IX.

Porphyrins vary in the type and sequence of side-chains on the tetrapyrrole ring. Many are derivatives of **Protoporphyrin IX**, which has the particular sequence shown above. Biologically active forms usually have a **metal ion** in the center, are associated with a **protein**, and take part in enzymic activities.

Heme: **iron protoporphyrin**, isomer **IX**. It occurs in:
> *hemoglobin* and *myoglobin* (O_2 transport and storage)
> *cytochromes* (electron transport)
> *catalase* (simple breakdown of H_2O_2)
> *peroxidases* (use H_2O_2 to oxidize an organic substrate).

Hemin, Hematin: terms which refer to the oxidized (Fe^{+++}) form of heme.

Tetrapyrroles (not heme) are also part of:
> chlorophyll
> vitamin B_{12}.

II. SYNTHESIS

ALA synthase is the first enzyme in the chain of heme synthesis:

$$
\begin{array}{ccccc}
\underset{\text{Succinyl CoA}}{\overset{\displaystyle \text{COOH}\quad\text{O}}{\underset{|}{\text{CH}_2\text{-CH}_2\text{-}}\overset{||}{\text{C}}\text{-SCoA}}} & + & \underset{\text{Glycine}}{\overset{\displaystyle \text{COOH}}{\underset{|}{\text{CH}_2\text{-NH}_2}}} & \xrightarrow[\substack{\textit{ALA}\\ \textit{Synthase}}]{\substack{\text{pyridoxal}\\ \text{phosphate}}} & \underset{\substack{\delta\text{-amino-}\\ \text{levulinic acid (ALA)}}}{\overset{\displaystyle \text{COOH}\quad\text{O}}{\underset{|}{\text{CH}_2\text{-CH}_2\text{-}}\overset{||}{\text{C}}\text{-CH}_2\text{-NH}_2}} \ + \ \text{CO}_2
\end{array}
$$

$+ \ \text{CoASH}$

An important control point, *ALA synthase* is rate-limiting and inhibited by the end product of the sequence, heme (or derivatives such as hemin [ferriheme]). Heme (and hemoglobin) synthesis is stimulated by **erythropoietin**, a glycoprotein.

Two ALA are then condensed to **porphobilinogen** (PBG), a monopyrrole. Four of these are joined, under the influence of two enzymes, *PBG deaminase* (older term, *Uroporphyrinogen III synthetase*) and *Uroporphyrinogen III cosynthase* to produce one particular isomeric cyclic tetrapyrrole, a porphyrinogen (**Uroporphyrinogen III**). This is then trimmed through several decarboxylations, giving rise to the specific sequence of methyl, vinyl, and propionic acid side-groups that is to be found in heme. The colorless product is next oxidized with NADPH and molecular oxygen to produce the colored protoporphyrin. Finally, **iron** is inserted by the enzyme *ferrochelatase*.

Heme synthesis occurs in all aerobic cells, but especially in blood-forming tissues. In man 80% of the body's heme is in **hemoglobin**, which reversibly binds and transports **oxygen**. Carbon monoxide competes strongly at the binding site with O_2, accounting for its toxicity.

III. METABOLISM

Heme is degraded in the reticuloendothelial system of liver, spleen and bone marrow. The metabolism involves a number of stages:

1. Ring opening by the complex enzyme heme oxygenase (with NADPH), to form **biliverdin**, a linear tetrapyrrole;

2. Reduction of biliverdin to produce **bilirubin**, an orange-yellow pigment which is insoluble in water;

3. Transport of the insoluble bilirubin as a complex with **serum albumin**;

4. Uptake by liver cells as a complex with **ligandin**, a binding protein;

5. Conjugation with **glucuronic acid**; this is obligatory for bilirubin to be secreted into bile:

$$
\text{Bilirubin} \ + \ \text{UDP Glucuronic acid} \ \xrightarrow[\textit{transferase}]{\textit{UDP-glucuronyl}} \ \text{Bilirubin diglucuronide}
$$

6. Excretion of the diglucuronide via bile through the intestinal tract.

Bilirubin is the major **bile pigment** in man. Bacterial metabolism of bilirubin in the bowel leads to formation of a complex mixture of pigments, including **urobilinogen**, a colorless substance which can readily undergo oxidation to **urobilin**. Some of these undergo an enterohepatic circulation and may also be found in urine. Thus urinary urobilin indicates that some degree of biliary excretion of bilirubin is occurring.

IV. DISORDERS OF PORPHYRIN AND HEME METABOLISM

A. Synthesis

These diseases usually take the form of overproduction and excretion of precursors (ALA, PBG) or porphyrin intermediates. Urine may be red. Called **porphyrias**, these conditions are usually inherited, but can sometimes be caused by toxic agents such as lead or some pesticides. They may be accompanied by photosensitivity of the skin, intermittent pain, muscular paralysis and psychic disturbances. ("Mad King George III" was not mad, but probably had one of the heritable porphyrias, see below). **Hematin** is useful in treatment because it inhibits ALA synthase.

B. Metabolism

Problems can occur at many points; build-up of bilirubin in blood and tissues causes a visible yellow color called **jaundice** (Table 6-1). A mild form often occurs in the newborn, due to a temporary underdevelopment of the conjugation and excretion process. Severe and prolonged accumulation of unconjugated bilirubin in the newborn is dangerous because unconjugated bilirubin can penetrate nervous tissue, concentrate in brain cells (**kernicterus**), and cause mental retardation.

Area of Defect in Bilirubin Metabolism	Cause	Main Form of Bilirubin Found
overproduction	hemolytic disorders	Unconjugated
uptake, conjugation and secretion by liver	underdevelopment in neonatal hyperbilirubinemia	Unconjugated
general hepatic deficiency	hepatitis, cirrhosis	Variable
biliary transport	blockage of bile duct by gallstone or cancer of pancreas	Conjugated
various	heritable defects in metabolism and transport, see below	

Table 6-1. Major Causes of Jaundice.

C. Genetic Defects

1. The Porphyrias

    Erythropoietic protoporphyria (EPP)
    Molecular defect: Deficient *ferrochelatase*.
    Pathway affected: Last step of heme synthesis.
    Expression: Increased protoporphyrin (plasma and feces).
    Genetics: Autosomal dominant.

    Acute intermittent porphyria (AIP)
    Molecular defect: Deficient *porphobilinogen deaminase*.
    Pathway affected: porphobilinogen → hydroxymethylbilane.
    Expression: Increased δ-aminolevulinic acid and PBG.
    Genetics: Autosomal dominant.

Other Hereditary Porphyrias
    Hereditary coproporphyria (HCP)
    Porphyria cutanea tarda (PCT)
    Variegate porphyria

2.   Bilirubin Metabolism: Hyperbilirubinemia

Crigler-Najjar Syndrome, Type 1
    Molecular defect: absent hepatic *bilirubin : UDP-glucuronyl transferase.*
    Pathway affected: bilirubin → bilirubin glucuronide.
    Expression: Increased unconjugated bilirubin.
    Genetics: Autosomal recessive.
    Treatment: Exchange transfusion.

Gilbert Syndrome (benign condition)
    Molecular defect: impaired hepatic uptake and conjugation of bilirubin.
    Pathway affected: bilirubin → bilirubin glucuronide.
    Expression: mild unconjugated hyperbilirubinemia.
    Genetics: Autosomal dominant?

V.   REVIEW QUESTIONS ON PORPHYRINS

MATCHING: For each set of questions, choose the ONE BEST answer from the list of lettered options above it. An answer may be used one or more times, or not at all.

Questions 1 - 6:

    A. "Direct" bilirubin in serum is elevated
    B. "Indirect" bilirubin in serum is elevated
    C. BOTH may be elevated
    D. NEITHER is elevated

1. Hepatitis.

2. Neonatal jaundice.

3. Acute intermittent porphyria.

4. Rh blood disease.

5. Cholelithiasis.

6. Crigler-Najjar Syndrome.

DIRECTIONS: For each of the following multiple-choice questions (7 - 22), choose the ONE BEST answer.

7. Bilirubin diglucuronide is

A. elevated in neonatal hyperbilirubinemia
B. usually found in the bile duct
C. lipid-soluble
D. elevated in hemolytic jaundice
E. normally excreted mainly in the urine.

8. Catalase

A. is a heme protein
B. is found in peroxisomes
C. breaks down hydrogen peroxide
D. produces oxygen as a product
E. all of the above are correct.

9. Unconjugated bilirubin is

A. elevated in neonatal hyperbilirubinemia
B. usually found in the bile duct
C. normally excreted in the urine
D. water-soluble
E. not involved in the development of kernicterus.

10. In porphyrin synthesis, the committed step is

A. condensation of 2 PBG (porphobilinogen)
B. condensation of glycine and succinyl CoA
C. isomerization of isopentenyl-PP to dimethyl-allyl-PP
D. condensation of 2 ALA (δ-aminolevulinic acid)
E. formation of uroporphyrinogen III.

11. The enzyme responsible for conjugation of bilirubin, prior to secretion into bile, is

A. ALA synthase
B. PBG deaminase
C. Uroporphyrinogen cosynthase
D. glucuronyl transferase
E. ferrochelatase.

12. The function of erythropoietin is to

A. promote synthesis of hemoglobin
B. prevent pernicious anemia
C. prevent thalassemia
D. promote binding of oxygen to hemoglobin
E. regulate reduction of methemoglobin.

13. Porphyrins are made from:

A. succinic acid and glycine
B. lysine and proline
C. alanine and glutamic acid
D. acetyl CoA and oxaloacetate
E. glycerol and fatty acid.

14. The first step in heme synthesis requires

A. thiamin
B. riboflavin
C. niacin
D. pyridoxal phosphate
E. cobalamin.

15. Which of the following does NOT contain an iron-porphyrin?

A. carboxyhemoglobin
B. catalase
C. peroxidase
D. myoglobin
E. ferredoxin.

16. Porphyria is associated with

A. phenylalanine-ammonia lyase deficiency
B. hypoxanthine-guanine phosphoribosyl transferase deficiency
C. overproduction of δ-aminolevulinic acid
D. bilirubin glucuronyl transferase deficiency
E. ALA synthetase deficiency.

17. A consequence of bile duct obstruction is

A. increased conjugated bile pigments in the serum
B. decreased excretion of cholesterol
C. decreased bilirubin glucuronides in the feces
D. increased levels of fecal lipids
E. all of the above.

18. Peroxidase

A. is a heme protein
B. is found in peroxisomes
C. breaks down hydrogen peroxide
D. oxidizes organic substrates
E. all of the above are correct.

19. Bile pigments can be produced from:

A. cholic acid
B. heme
C. triglycerides
D. cholesterol
E. phospholipids.

20. Urobilinogen can NOT be derived from

A. hemoglobin
B. catalase
C. flavoprotein
D. cytochrome c
E. peroxidase.

21. Urobilinogen is formed in the

A. spleen
B. liver
C. bone marrow
D. bowel
E. kidney.

22. The last step in heme synthesis is catalyzed by:

A. Ferrochelatase
B. Uroporphyrinogen cosynthase
C. ALA synthase
D. UDP-glucuronyl transferase
E. heme oxygenase

VI. ANSWERS TO QUESTIONS ON PORPHYRINS

1.	C		12.	A	
2.	B		13.	A	
3.	D		14.	D	
4.	B		15.	E	
5.	A		16.	C	
6.	B		17.	E	
7.	B		18.	E	
8.	E		19.	B	
9.	A		20.	C	
10.	B		21.	D	
11.	D		22.	A	

7. LIPIDS
Chi-Sun Wang

I. CLASSIFICATION OF LIPIDS

A. Major Lipid Classes

Fatty acids: long-chain monocarboxylic acids.

Acylglycerols: triacylglycerols, diacylglycerols and monoacylglycerols.

Phospholipids: phosphoglycerides and sphingomyelin.

Glycosphingolipids: cerebroside (ceramide monosaccharide), cerebroside sulfate, ceramide oligosaccharide and ganglioside. The common structural component of glycosphingolipids is ceramide (N-acylsphingosine). They are similar to sphingomyelin as derivatives of a ceramide, but do not contain phosphorus and the additional nitrogenous base.

Other lipids: sterols, terpenes, waxes, aliphatic alcohols.

Lipoproteins: lipids combined with other classes of compounds.

B. Derived lipids

These are molecules, soluble in lipid solvents, that are produced by hydrolysis of natural lipids.

II. FUNCTIONS OF LIPIDS

A. Structural Component

All cellular membranes, including myelin, consist of a lipid bilayer. Membranes function as permeability barriers and in many other ways. See Chapter 9.

B. Enzyme Cofactors

Several enzymes require lipids for their activity. Examples include phospholipid in the blood clotting cascade, coenzyme A, etc.

C. Energy Storage

Energy storage is the major function of triacylglycerols which are largely found in adipose tissue. These lipids do not require hydration and yield 9 Kcal/g of energy on complete combustion, compared to 4 Kcal/g for carbohydrates and 4 Kcal/g for proteins. This is because the C-H / O ratio is higher.

D. Hormones and Vitamins

Prostaglandins: arachidonic acid is a precursor for the biosynthesis of prostaglandins.

Steroid Hormones: estradiol, testosterone, cortisol, and many other gonadal and adrenal cortical hormones produced from cholesterol.

Fat-soluble Vitamins: A, D, E, K.

E. Aid in Fat Digestion: *Components of Bile*

F. Insulation: *Subcutaneous Adipose Tissue*

G. Thermogenesis: *Brown Adipose Tissue*

III. DIGESTION AND ABSORPTION OF LIPIDS

Most dietary lipid is in the form of triacylglycerols and is digested in the small intestine. The emulsification of lipid droplets by **bile salts** allows a large increase in the surface area of the droplets. Lipase produced by the pancreas catalyzes the hydrolysis of triacylglycerols to fatty acids and monoacylglycerols, which form **micelles**. These micelles also contain bile salts, cholesterol and fat-soluble vitamins. The micelles then migrate by diffusion to intestinal mucosa and enter intestinal epithelial cells, where free fatty acids and monoacylglycerols are re-esterified to form triacylglycerols. Triacylglycerol, cholesterol, cholesterol ester, phospholipid, and specific proteins are assembled into **chylomicron** particles which subsequently enter the lymphatics for transport to the thoracic duct and the bloodstream.

IV. FATTY ACIDS

A. Chemistry of Fatty Acids

Saturated	Nbr of Carbons	Structure
Acetic	2	CH_3COOH
Propionic	3	CH_3CH_2COOH
Butyric	4	$CH_3(CH_2)_2COOH$
Hexanoic	6	$CH_3(CH_2)_4COOH$
Octanoic	8	$CH_3(CH_2)_6COOH$
Decanoic	10	$CH_3(CH_2)_8COOH$
Lauric	12	$CH_3(CH_2)_{10}COOH$
Myristic	14	$CH_3(CH_2)_{12}COOH$
Palmitic	16	$CH_3(CH_2)_{14}COOH$
Stearic	18	$CH_3(CH_2)_{16}COOH$

Unsaturated	Nbr of Double Bonds	Structure
Oleic	1	$cis\text{-}\Delta^9$
Linoleic	2	$cis,cis\text{-}\Delta^{9,12}$
Linolenic	3	all $cis\text{-}\Delta^{9,12,15}$
Arachidonic	4	all $cis\text{-}\Delta^{5,8,11,14}$

Table 7-1. Some Common Fatty Acids.

Comments on Fatty Acids. $Cis\text{-}\Delta^9$ for oleic acid means that the double bond is between the 9th and 10th carbons (starting with the carboxyl carbon as number 1); the double bond has the *cis* geometric configuration:

$$
\begin{array}{ccc}
H & & H \\
\backslash & & / \\
C & = & C \\
/ & & \backslash \\
CH_3(CH_2)_7 & & (CH_2)_7COOH
\end{array}
$$

The most common fatty acids found in mammals are straight-chain, unsubstituted, monocarboxylic acids and have an even number of carbons, usually 12 - 20 (Table 7-1). When a fatty acid contains more than one double bond, it is polyunsaturated. The polyunsaturated fatty acids linoleic and linolenic are essential in the diet of mammals.

B. β-Oxidation of Fatty Acids

Steps in β-Oxidation (Figure 7-1). After initial **activation** of a fatty acid in the cytoplasm (enzymatic step 1 in the figure), the entrance of acyl-CoA into mitochondria requires the **carnitine transport** system. The β-oxidation of acyl-CoA then consists of a repeated series of four steps: **dehydrogenation, hydration, dehydrogenation**, and **thiolytic cleavage** (steps 2-5 in the figure). The last step in each series releases a 2-carbon fragment as acetyl CoA.

Cofactor Requirement. ATP (for activation), carnitine (for entrance into mitochondria), NAD^+, FAD, and CoA.

Stoichiometry. The stoichiometry for the oxidation of palmitate ($C_{16}H_{32}O_2$) can be calculated as shown below:

Palmitate		
7 turns of β-oxidation (7 NAD^+, 7 FAD)	+ 35	ATP
8 acetyl groups	+ 96	ATP
1 ATP used for initial activation	− 2	ATP
(but cleavage of PP_i costs an extra ~P)		
Net:	129	ATP

Oxidation of unsaturated fatty acids requires cis-Δ^3 → trans-Δ^2-enoyl-CoA isomerase.

C. Catabolism of Odd-numbered-carbon Fatty Acids

Oxidation of an odd-numbered-carbon fatty acid yields successive molecules of acetyl-CoA, ending with 1 of **propionyl-CoA**. The major pathway for the metabolism of propionyl-CoA is summarized in the three reactions shown below (Figure 7-2):

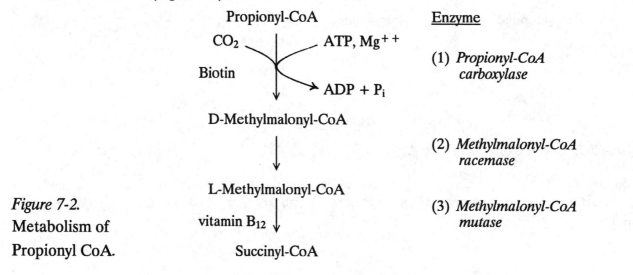

Figure 7-2. Metabolism of Propionyl CoA.

In these enzyme reactions, **biotin** is a cofactor for propionyl-CoA carboxylase and **vitamin B_{12}** is a cofactor for methylmalonyl-CoA mutase. Succinyl-CoA can then enter into the TCA cycle.

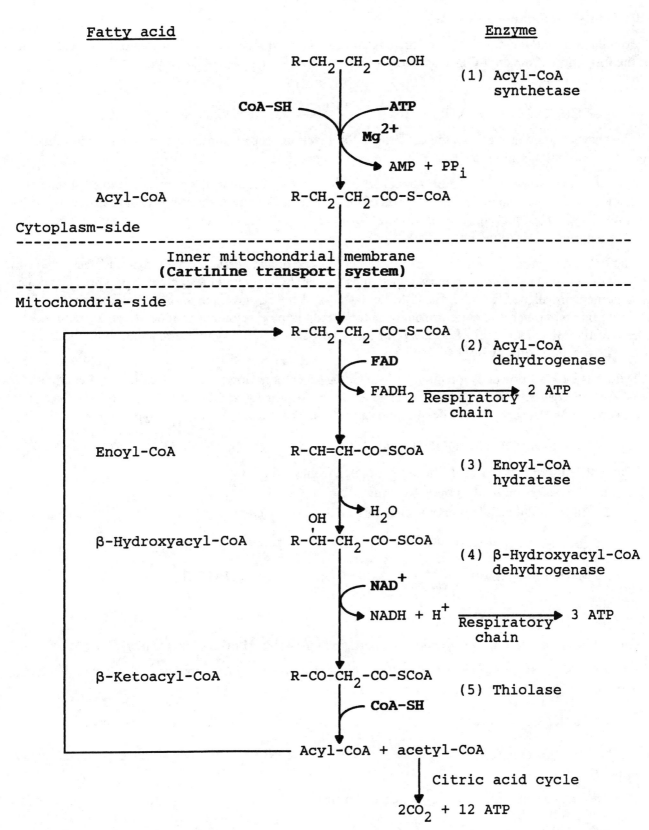

Figure 7-1. β-Oxidation of Fatty Acids in Mitochondria.

D. Biosynthesis of Saturated Fatty Acids

Formation of malonyl-CoA. The major *de novo* route is in cytoplasm. The first step of fatty acid synthesis is the formation of **malonyl-CoA** from acetyl-CoA catalyzed by *acetyl-CoA carboxylase*.

$$CH_3\text{-}\overset{\overset{\displaystyle O}{\|}}{C}\text{-S-CoA} + ATP + HCO_3^- \xrightarrow{\text{Biotin}} \text{Malonyl-CoA} + ADP + P_i$$

The enzyme contains biotin as the cofactor in the reaction. Biotin-containing enzymes cannot function if absorption of biotin is inhibited by **avidin**, a protein in egg white.

Fatty acid synthase (FAS). There appear to be two types of fatty acid synthase systems. In bacteria, plants, and lower forms, the individual enzymes of the system are separate and the acyl radicals are in combination with acyl carrier protein (ACP). In yeast, mammals, and birds, the enzymes are combined as a **multienzyme complex** that is found in cytoplasm.

The fatty acid synthase complex of mammals is composed of a dimer with a monomeric molecular weight of 267,000. In each monomer all 6 enzyme activities and an ACP reside in a single polypeptide chain. The two monomers are aligned in a cyclic head-to-tail fashion. The thiol group in $4'$-phosphopantetheine (Pan) of ACP on one monomer is in close proximity to the thiol group of a cysteine residue attached to β-ketoacyl synthase of the other monomer. Because both thiols participate in the synthase reaction, only the dimer is active. The reaction pathway is shown in Figure 7-3.

Synthesis by FAS proceeds for fatty acids of up to 16 carbon atoms. The control step is catalyzed by *acetyl-CoA carboxylase*. The enzyme is inducible, **activated by citrate**, and **inhibited by fatty acyl CoA**. The resulting malonyl-CoA carbons are added to the carboxyl end of the fatty acid and CO_2 is released:

$$R\text{-CO-ACP} + {}^-OOC\overset{*}{C}H_2\overset{*}{C}O\text{-CoA} \longrightarrow RCO\overset{*}{C}H_2\overset{*}{C}O\text{-ACP} + CO_2$$

The synthesis requires **NADPH** which is generated from:
1. The **hexose monophosphate pathway**;
2. An NADP-dependent **malate enzyme** ("malic enzyme"):

$$\text{Oxaloacetate (cytosol)} + NADH + H^+ \xrightarrow[\text{dehydrogenase}]{\text{malate}} \text{malate} + NAD^+$$

$$\text{Malate} + NADP^+ \xrightarrow[\text{enzyme}]{\text{malic}} \text{pyruvate} + CO_2 + NADPH + H^+$$

E. Fatty Acid Synthesis from Carbohydrate

Lipogenesis from carbohydrate requires the participation of the **mitochondrion** (Figure 7-4).

Transport of acetyl-CoA to the cytoplasm is mediated *via* **citrate**, whose formation in mitochondria is catalyzed by *citrate synthase*. Citrate is then converted to acetyl-CoA in cytoplasm by *ATP-citrate lyase*:

$$\text{Citrate} + ATP + CoA \xrightarrow[\text{lyase}]{\text{ATP-citrate}} \text{acetyl-CoA} + ADP + P_i + \text{Oxaloacetate}$$

The synthesis of this inducible enzyme is increased in caloric excess. The acetyl-CoA is then utilized for biosynthesis of fatty acids.

Reducing equivalents are needed, generated from the hexose monophosphate pathway and "malic enzyme."

Animals cannot convert fatty acids into glucose because pyruvate → acetyl-CoA is irreversible.

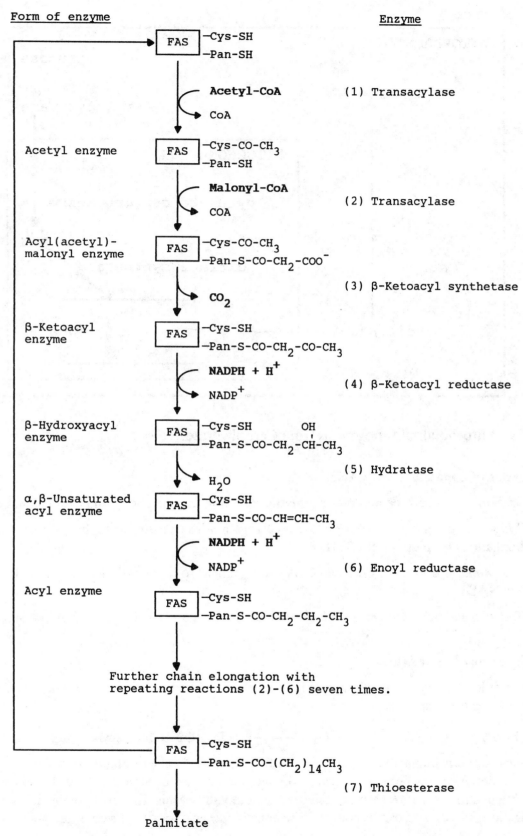

Figure 7-3. Reactions Carried Out by Fatty Acid Synthase.

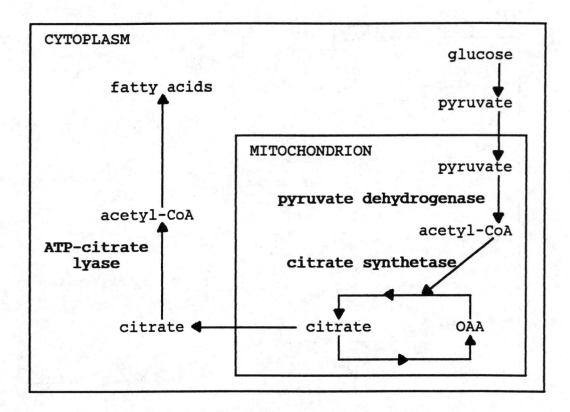

Figure 7-4. Mitochondrial Lipogenesis from Carbohydrate.

F. Elongation of Long Chain Fatty Acids

Elongation occurs in three areas — cytoplasm, mitochondria and endoplasmic reticulum:

1. *Cytoplasm* contains a system for fatty acid elongation which is similar to the one used for *de novo* synthesis. Elongation also requires NADPH.

2. *Mitochondria* use acetyl-CoA and CoA esters, similar to a reversal of β-oxidation, but require both NADH and NADPH.

3. *Endoplasmic reticulum (microsomes)* use malonyl CoA and CoA ester, not ACP-esters. NADPH is required.

G. Desaturation of Fatty Acids.

Desaturation occurs in endoplasmic reticulum. *NADPH is required.* Both the fatty acid and NADPH are oxidized by molecular oxygen:

$$\text{Fatty acyl-CoA} + \text{NADPH} + \text{H}^+ + \text{O}_2 \longrightarrow \text{Unsaturated fatty acyl-CoA} + \text{NADP}^+ + 2\,\text{H}_2\text{O}$$

All desaturations in mammalian systems are **farther than 6 carbon atoms** from the methyl end of the fatty acid. Therefore, any fatty acid having a double bond closer to the methyl end than 6 carbons must either be obtained in the diet or be produced from a dietary fatty acid. Thus by a combination of desaturation and elongation, linoleic acid can be converted to arachidonic acid for conversion to prostaglandins.

Double bonds are all cis, although *trans* are present occasionally in fats from plant sources. Partially hydrogenated margarine may also contain *trans* fatty acids which result from the manufacturing process.

H. Prostaglandins

The polyunsaturated C-20 fatty acids give rise to physiologically and pharmacologically active acid derivatives known as **eicosanoids**: prostaglandins (PG), prostacyclins (PGI), thromboxanes (TX), and leukotrienes (LT). Eicosanoids can be put into three groups, based on the polyunsaturated acid from which they are derived. The groups, their biosynthetic origins, and some examples are shown below (Figure 7-5).

Group and Parent Acid	Examples	
Group (1)		
	PGE_1	
	PGF_1	Prostanoids
8,11,14-Eicosatrienoate	TXA_1	
20:3(8,11,14)		
	LTA_3	Leukotrienes
Group (2)		
	PGE_2	
	PGF_2	Prostanoids
5,8,11,14-Eicosatetraenoate	PGI_2	
20:4(5,8,11,14)	TXA_2	
(Arachidonic Acid)		
	LTA_4	Leukotrienes
Group (3)		
	PGE_3	
	PGF_3	Prostanoids
5,8,11,14,17-Eicosapentaenoate	PGI_3	
20:5(5,8,11,14,17)	TXA_3	
	LTA_5	Leukotrienes

Figure 7-5. Relationships Among the Eicosanoids.

The major classes of prostaglandins (as they are known collectively) are designated with a numerical subscript which denotes the **number of double bonds** in the molecule. Because of a cyclization reaction, prostanoids have 2 fewer double bonds than the parent fatty acid, while leukotrienes have the same number.

Two major pathways are followed in the synthesis of the eicosanoids:

(1) The **cyclooxygenase pathway** produces the prostanoids. It includes prostaglandin endoperoxide which has two enzyme activities, *cyclooxygenase* and *peroxidase*, and consumes two molecules of O_2. **Aspirin** was found to inhibit the cyclooxygenase reaction. By this pathway, **thromboxanes** are synthesized in platelets and upon release cause **vasoconstriction** and **platelet aggregation**. **Prostacyclins** (eg., PGI_2) are produced by blood vessel walls and are potent **inhibitors of platelet aggregation**.

(2) The **lipoxygenase pathway** produces **leukotrienes**, a family of conjugated trienes. The slow-reacting substrate of anaphylaxis is a mixture of LTC_4, LTD_4, LTE_4. These leukotrienes also cause vascular permeability and are important in **inflammatory** and immediate **hypersensitivity reactions**.

V. KETONE BODIES

A. Synthesis of Ketone Bodies

Ketogenesis is a hepatic process (Figure 7-6). In liver mitochondria, fatty acids are oxidized to **acetyl CoA**, a portion of which is oxidized via the TCA cycle to CO_2 and a portion of which is converted via **3-hydroxy-3-methylglutaryl-CoA** (HMG-CoA) to the ketone bodies, acetoacetate and β-hydroxybutyrate.

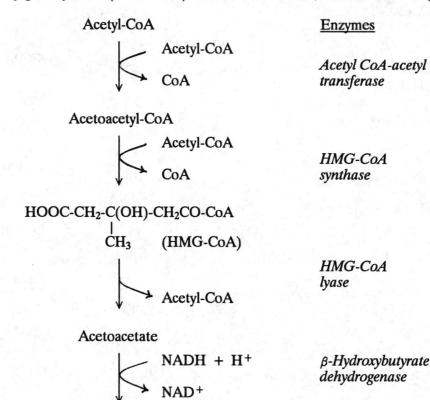

Figure 7-6. Hepatic Ketogenesis.

B. Metabolism and Regulation.

Utilization. Ketone bodies are utilized primarily by muscle.

Starvation. Liver glycogen in the rat is depleted by about 70% after a 12-hour fast and is almost gone after 24 hours of starvation. Acetyl-CoA derived from fatty acid β-oxidation is channeled into ketogenesis. The synthetic pathways for carbohydrates and fatty acids are inhibited. Normally, adult brain depends entirely on glucose for its energy. Under conditions of long-term starvation (more than 3 days), the brain derives a considerable part of its energy from ketone bodies.

Insulin deficiency. In regard to lipid metabolism, diabetes and starvation resemble each other. Fatty acids mobilized from adipose tissue raise the level of acetyl-CoA in the liver and promote **ketogenesis**.

Consequences of ketosis. Ketone bodies are excreted in large part as the sodium salts. Depletion of Na^+ and other cations leads to **acidosis** (ketoacidosis).

VI. TRIACYLGLYCEROLS

A. Synthesis *de novo* is mainly in liver and adipose tissue (Figure 7-7).

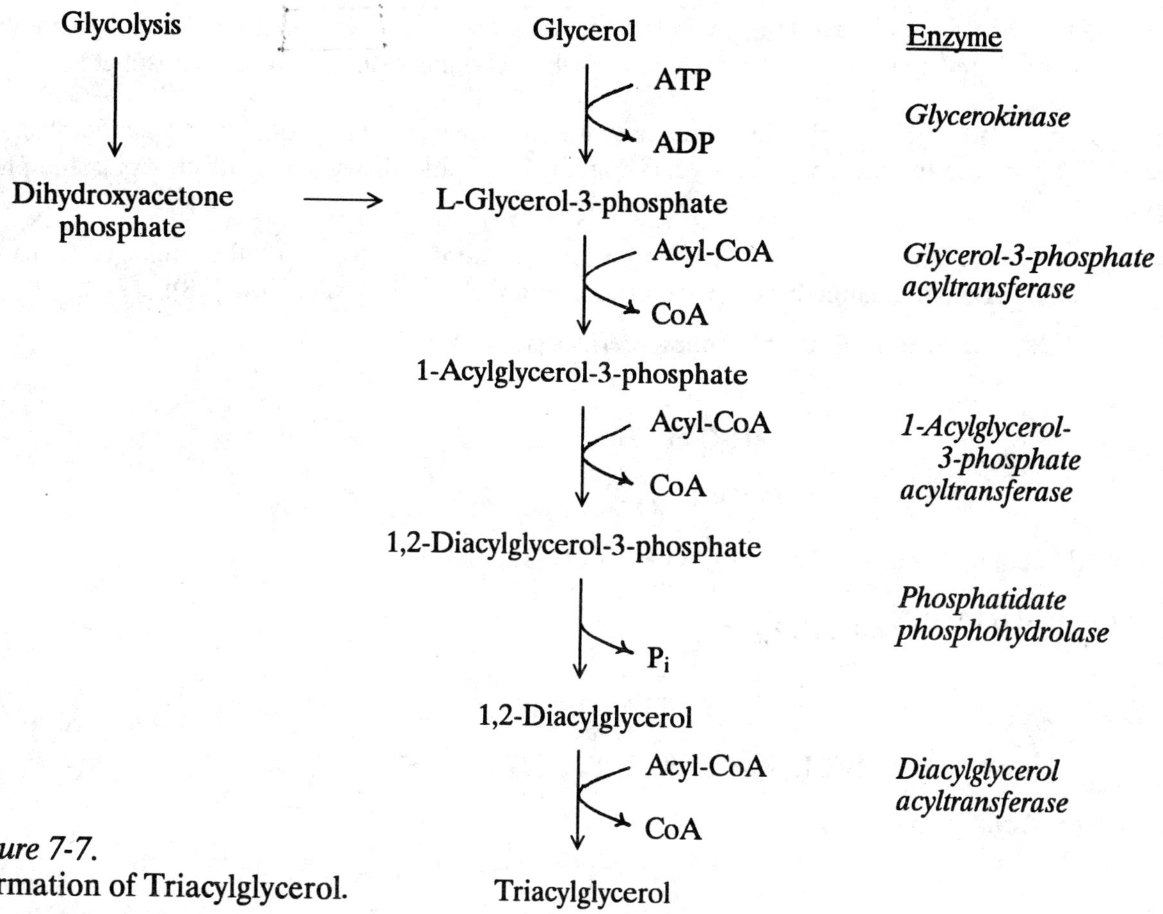

Glycolysis	Glycerol	<u>Enzyme</u>
↓	⤷ ATP	
	⤷ ADP	*Glycerokinase*
Dihydroxyacetone phosphate →	L-Glycerol-3-phosphate	
	⤷ Acyl-CoA	*Glycerol-3-phosphate acyltransferase*
	⤷ CoA	
	1-Acylglycerol-3-phosphate	
	⤷ Acyl-CoA	*1-Acylglycerol-3-phosphate acyltransferase*
	⤷ CoA	
	1,2-Diacylglycerol-3-phosphate	
	⤷ P_i	*Phosphatidate phosphohydrolase*
	1,2-Diacylglycerol	
	⤷ Acyl-CoA	*Diacylglycerol acyltransferase*
	⤷ CoA	
	Triacylglycerol	

Figure 7-7.
Formation of Triacylglycerol.

B. Utilization

Digestion. Dietary triacylglycerols are hydrolyzed in the intestine and resynthesized by intestinal cells, then transferred to the lymph as **chylomicrons**. Secretion of chylomicrons requires intestinal protein synthesis.

Transport. Triacylglycerols produced in liver are released as **very low density lipoproteins** (VLDL).

Lipoprotein lipase is an enzyme bound to capillary endothelium. Peripheral tissues require it in order to utilize the triacylglycerol from triacylglycerol-containing lipoproteins. (See also section IX F, page 119).

Hormone-sensitive lipase (HSL) hydrolyzes triacylglycerols to fatty acids and glycerol during fat mobilization in adipose tissue. (See also section IX G, page 119). Because adipose tissue has no glycerol kinase, glycerol is not re-utilized there but is metabolized predominantly by the liver.

Regulation. Fatty acid taken up by the liver undergoes (1) oxidation, or (2) esterification to triacylglycerols. In the normal fed condition, the liver has plenty of carbohydrate to be oxidized for energy, therefore (2) > (1). In the fasting state, fatty acid oxidation proceeds at a higher rate, therefore (1) > (2).

The esterification pathway (2) is never saturated, therefore the shift to oxidation in fasting is not because esterification has reached a maximum; it is because the oxidative pathway is turned on.

VII. PHOSPHOLIPIDS

A. Phosphoglycerides

Structure. Phosphoglycerides have long-chain fatty acids esterified to glycerol in the 1 and 2 positions, a phosphate in position 3, and a nitrogen-containing base, such as choline, esterified to the phosphate.

Phospholipases. There are several phospholipases which hydrolyze phospholipids at specific places in the molecule. They are useful in investigations requiring highly specific cleavage of phospholipids and phosphate esters.

Plasmalogens. These phosphoglycerols contain an α,β-unsaturated alcohol in ether linkage. Thus the plasmalogens containing ethanolamine have the general structure shown below (Figure 7-8).

Synthesis. The synthetic pathway for phosphoglycerides is shown in Figure 7-9.

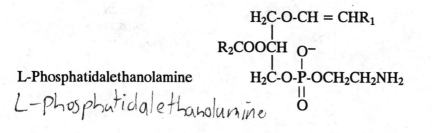

$$H_2C\text{-}O\text{-}CH = CHR_1$$
$$R_2COOCH$$

L-Phosphatidalethanolamine

$$H_2C\text{-}O\text{-}\overset{O^-}{\underset{\underset{O}{\parallel}}{P}}\text{-}OCH_2CH_2NH_2$$

L-phosphatidalethanolamine

Ceramide (N-Acyl-Sphingosine)

$$CH_3(CH_2)_{12}\text{-}CH = CH\text{-}\overset{OH}{\underset{\underset{\overset{\parallel}{O}}{HN\text{-}C\text{-}R}}{CH}}\text{-}CH\text{-}CH_2OH$$

Sphingomyelin

$$CH_3(CH_2)_{12}\text{-}CH = CH\text{-}\overset{OH}{\underset{\underset{\overset{\parallel}{O}}{HN\text{-}C\text{-}R}}{CH}}\text{-}CH\text{-}CH_2\text{-}O\text{-}\overset{O^-}{\underset{\underset{O}{\parallel}}{P}}\text{-}O\text{-}CH_2\text{-}CH_2\text{-}N^+(CH_3)_3$$

Figure 7-8. Structures of Some Polar Lipids.

B. Sphingomyelin

Structure: Ceramide and choline in a phosphodiester linkage (Figure 7-8).

Synthesis:

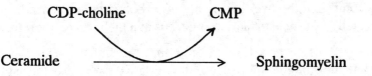

CDP-choline   CMP

Ceramide ——————————→ Sphingomyelin

Figure 7-9.
Synthetic Pathway for Phosphoglycerides.

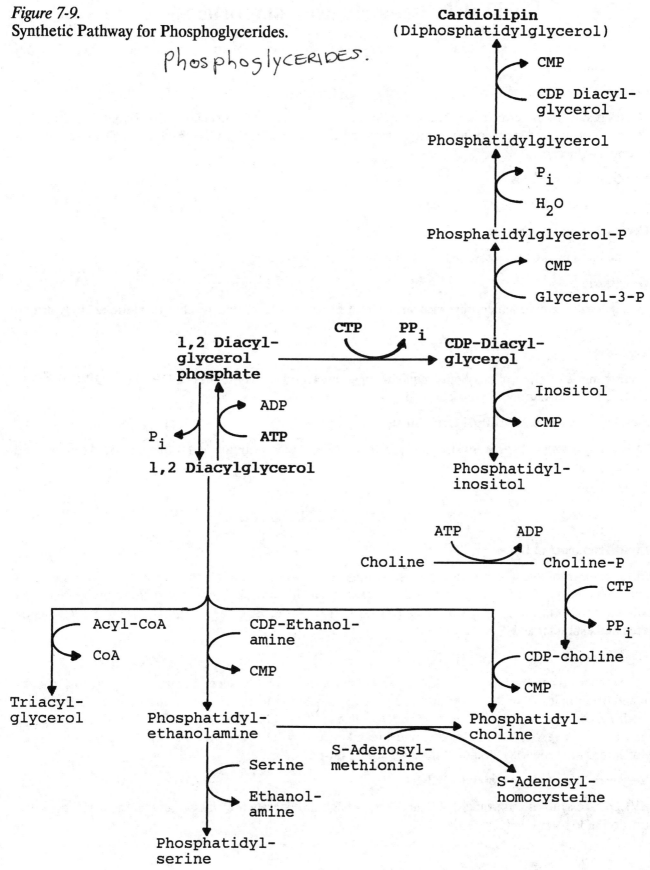

VIII. GLYCOSPHINGOLIPIDS (GLYCOLIPIDS).

Glycosphingolipids are the same as sphingomyelin in that they contain ceramide but no glycerol. They do not contain phosphate.

A. Cerebroside: Ceramide-hexose

Cerebroside, found primarily in brain, consists of a hexose linked to ceramide. On complete hydrolysis it yields one molecule of sphingosine, a fatty acid, and a hexose which is most frequently D-galactose, less commonly D-glucose.

B. Cerebroside Sulfatides

These are sulfate esters of galactocerebroside.

C. Ceramide Oligosaccharides

Designated by the terms ceramide disaccharide, ceramide trisaccharide, etc.

D. Gangliosides

Gangliosides are ceramide oligosaccharides that contain at least one residue of **sialic acid** (N-acetyl-neuraminic acid).

E. Synthesis

Biosynthesis of glycosphingolipids requires **sugar nucleotides**. These are: UDP-Glc, UDP-Gal, UDP-GlcNAc, UDP-GalNAc and CMP-sialic acid.

F. Metabolic Blocks in Glycolipid Catabolism

Several **lysosomal disorders** exist, in which various glycolipids accumulate ("Lipid storage diseases," see Section X, page 120).

IX. LIPOPROTEINS

A. Lipid Transport in Blood

Lipids are insoluble in water and so must be transported as lipoproteins. Lipoproteins are classified according to increasing density: chylomicrons, very low density lipoproteins, low density lipoproteins, and high density lipoproteins. Nine apolipoproteins (ApoA-I, A-II, A-IV, B, C-I, C-II, C-III, D and E) have been isolated and characterized.

B. Chylomicrons

Chylomicrons are the largest of the lipoprotein particles which transport dietary triacylglycerols, cholesterol and other lipids from the intestine to adipose tissue and the liver. The triacylglycerols in chylomicrons are hydrolyzed within a few minutes by *lipoprotein lipase*. The cholesterol-rich residual particles, known as **remnants**, retain ApoE but not ApoC, and are taken up by the liver by a receptor specific for ApoE. Chylomicrons are especially abundant in blood after a fat-rich meal.

C. Very Low Density Lipoproteins (VLDL)

VLDL are synthesized primarily by the liver. VLDL are degraded by *lipoprotein lipase*, resulting in the production of low density lipoproteins.

D. Low Density Lipoproteins (LDL)

Most LDL appears to be formed from VLDL, but there is evidence for some direct production by the liver. These lipoproteins are rich in **cholesterol ester** and are **atherogenic**.

Cells outside the liver and intestine obtain cholesterol from plasma LDL rather than by synthesizing it *de novo*. Uptake from blood is by a **receptor-mediated** process specific for ApoB. The steps in the uptake are: (1) LDL binds to regions of plasma membranes called "**coated pits.**" (2) The receptor-LDL complexes are internalized by **endocytosis**. (3) These vesicles subsequently fuse with lysosomes, are degraded, and the cholesterol is released within the cell.

The cholesterol content of cells that have an active pathway for uptake of LDL is regulated in two ways. First, the released cholesterol **suppresses** the activity of 3-hydroxy-3-methylglutaryl-CoA reductase (*HMG-CoA reductase*), thereby inhibiting *de novo* synthesis of cholesterol. Second, the number of LDL receptors is regulated with a feedback mechanism. When cholesterol is abundant inside the cell, new **LDL receptors are not synthesized** and the uptake of additional cholesterol from plasma LDL decreases.

The genetic defect in most cases of **familial hypercholesterolemia** is due to an absence or deficiency of the normal receptor for LDL.

E. High Density Lipoproteins (HDL)

HDL are produced by the liver. A major function of HDL is to act as a repository for apolipoproteins C and E that are required in the metabolism of chylomicrons and VLDL. HDL also are important in the transport of cholesterol from the periphery to the liver. Plasma LCAT (*lecithin : cholesterol acyl transferase*) is activated by ApoA-I of HDL and brings about esterification of cholesterol on the surface of HDL, resulting in its accumulation in the interior of the lipoprotein particles. This can create a concentration gradient and draw cholesterol from extrahepatic tissues for reverse cholesterol transport; thus **HDL is anti-atherogenic**.

F. Lipoprotein Lipase (LPL)

Triacylglycerols in chylomicrons and VLDL cannot be taken up intact by tissues but must first undergo hydrolysis by LPL, an enzyme situated on the capillary endothelium of extrahepatic tissues. The enzyme requires apolipoprotein C-II as cofactor.

G. Mobilization of Fat by Hormone-Sensitive Lipase (HSL)

The fat reserves of mammals are stored in **adipose tissue**. Fatty acids in triacylglycerols must first be hydrolyzed to free fatty acids before they can be transported from adipose tissue to other tissues. The hydrolysis of the triacylglycerols within adipose tissue is catalyzed by *hormone-sensitive lipase*. The released free fatty acids are transported in blood by the plasma protein, **albumin**.

The lipolysis of adipose tissue fat (fat mobilization) is stimulated by **glucagon and epinephrine**. Lipolysis is inhibited by **insulin**, which stimulates glucose uptake for lipogenesis, decreasing cAMP in fat cells.

H. Hypo- and Hyperlipoproteinemias

A number of heritable disorders of lipoprotein metabolism occur. (See Section X, page 120).

I. Fatty Liver.

Two main categories occur: The first is associated with a raised level of plasma free fatty acids. The second is usually due to a metabolic block in the production of plasma lipoproteins, which may be caused by a deficiency of choline. Orotic acid also may produce fatty liver, perhaps by inhibiting glycosylation of ApoB and thus interfering with the release of VLDL into plasma.

J. Hypolipidemic Drugs.

Compactin and *mevinolin*: inhibit the HMG-CoA reductase reaction.

Clofibrate and *gemfibrozil*: reduce hepatic synthesis of triacylglycerol.

Colestipol and *cholestyramine*: resins that bind bile salts and prevent reabsorption.

Probucol: enhances catabolism of LDL.

Nicotinic acid: reduces lipolysis in adipose tissue.

Neomycin: inhibits reabsorption of bile salts.

X. GENETIC DISEASES

A. Lipoprotein metabolism

Familial *lipoprotein lipase* deficiency
  Molecular defect: Deficient lipoprotein lipase
  Pathway affected: Hydrolysis of glyceride ester bonds.
  Diagnosis: Accumulation of chylomicron triglyceride.
  Genetics: Autosomal recessive.

Familial hypercholesterolemia
  Molecular defect: Abnormal LDL receptors; 3 mutations:
      1. no receptor binding of LDL
      2. reduced receptor binding of LDL
      3. no internalization of bound LDL
  Pathway affected: Cellular uptake and delivery of LDL to lysosome.
  Diagnosis: High cholesterol in plasma.
  Genetics: Autosomal dominant.

Familial *lecithin : cholesterol acyltransferase* (LCAT) deficiency
  Molecular defect: Deficient LCAT.
  Pathway affected: Cholesterol → cholesteryl ester.
  Diagnosis: Abnormal plasma lipoprotein pattern.
  Genetics: Autosomal recessive.

B. Sphingolipidoses

G_{M1}-gangliosidosis types 1, 2 and 3
  Molecular defect: Mutated *β-gangliosidase A*.
  Pathway affected: Ganglioside G_{M1} → G_{M2}
  Diagnosis: Reduced enzyme activity in leukocytes.
  Genetics: Autosomal recessive.

Tay-Sachs, G_{M2}-gangliosidosis type 1

    Molecular defect: Deficient *hexosaminidase A*; mutated α chain.

    Pathway affected: $G_{M2} \rightarrow$ N-acetyl-β-D-galactosamine.

    Diagnosis: Deficient Hex A in serum and leukocytes.

    Genetics: Autosomal recessive.

Sandhoff, G_{M2}-gangliosidosis type 2

    Molecular defect: Deficient *Hex A and B*; mutated β chain.

    Pathway affected: $G_{M2} \rightarrow$ N-acetyl-β-D-galactosamine.

    Diagnosis: Deficient activities of Hex A and Hex B in serum.

    Genetics: Autosomal recessive.

Gaucher, types 1, 2 and 3

    Molecular defect: Deficient *β-glucocerebrosidase*

    Pathway affected: Glucocerebroside $\rightarrow$ Glc

    Diagnosis: Reduced enzyme activity in leukocytes

    Genetics: Autosomal recessive (all three types).

Krabbe, globoid cell leukodystrophy

    Molecular defect: Deficient *galactocerebroside b-galactosidase*.

    Pathway affected: galactocerebroside $\rightarrow$ ceramide + Gal.

    Diagnosis: deficient enzyme activity in leukocytes.

    Genetics: Autosomal recessive.

Niemann-Pick, types I and II

    Molecular defect: Deficient *sphingomyelinase*.

    Pathway affected: Sphingomyelin $\rightarrow$ ceramide + phosphorylcholine.

    Diagnosis: Deficient enzyme activity in leukocytes

    Genetics: Autosomal recessive.

Metachromatic leukodystrophy

    Molecular defect: Deficient *arylsulfatase A*

    Pathway affected: Catabolism of sulfatides and sulfogalactoglycerolipids.

    Diagnosis: Deficient enzyme activity in leukocytes

    Genetics: Autosomal recessive.

Fabry

    Molecular defect: Deficient *α-galactosidase A*.

    Pathway affected: Gal-Gal-Glc-Cer $\rightarrow$ Gal-Glc-Cer + Gal.

    Diagnosis: Increased tissue Gal-Gal-Glc-Cer; deficient α-galactosidase A (plasma).

    Genetics: X-linked.

XI. REVIEW QUESTIONS ON LIPIDS

DIRECTIONS: For each of the following multiple-choice questions (1 - 55), choose the
ONE BEST answer.

---

1. Glucose catabolism through the pentose phosphate pathway stimulates fatty acid synthesis because it increases

A. acetyl-CoA.
B. NADH.
C. NADPH.
D. ATP.
E. glycogen.

2. Which of the following is NOT on the major route to β-hydroxybutyrate in liver?

A. acetyl-CoA.
B. malonyl-CoA.
C. acetoacetyl-CoA.
D. acetoacetate.
E. hydroxymethyl glutaryl-CoA.

3. Acetyl-CoA carboxylase and other biotin-containing enzymes have lowered activity after ingestion of dietary

A. citrate.
B. carnitine.
C. avidin.
D. lactalbumin.
E. cyanide.

4. β-Oxidation of fatty acids involves a sequence of four reactions repeated several times. The sequence is best described by:

A. oxidation - decarboxylation - dehydration - oxidation.
B. oxidation - dehydration - oxidation - thiolysis.
C. condensation - oxidation - dehydration - oxidation.
D. condensation - hydration - oxidation - thiolysis.
E. oxidation - hydration - oxidation - thiolysis.

5. In the lipogenesis of fatty acids from pyruvate, which of the following enzyme reactions occurs in cytoplasm?

A. pyruvate dehydrogenase.
B. citrate synthase.
C. propionyl-CoA carboxylase.
D. succinate dehydrogenase.
E. ATP-citrate lyase.

6. About how many grams of stearate must be metabolized to give an amount of energy equivalent to 50 g of glycogen?

A. 5
B. 20
C. 40
D. 80
E. 94

7. The cofactors common to both oxidation and synthesis of fatty acids include

A. NAD, NADP.
B. FAD, NADP.
C. NAD, CoA.
D. CoA, ATP.
E. NAD, ATP.

8. In the control of fatty acid metabolism,

A. when fuel is abundant fatty acid oxidation is enhanced.
B. high carbohydrate intake causes increased fatty acid synthesis.
C. high carbohydrate intake causes decreased fatty acid synthesis.
D. palmitoyl CoA has no effect on fatty acid synthesis.
E. none of the above is correct.

9. Increased urinary acetoacetate might result when liver

A. glycogen is normal.
B. glycogen is depleted.
C. NADPH concentration is high.
D. acetyl CoA concentration is low.
E. none of the above.

10. The role of hormone-sensitive lipase is to

A. hydrolyze the ester bonds in hormones.
B. hydrolyze dietary fat and the enzyme is stimulated by epinephrine.
C. mobilize fat from adipose tissue.
D. hydrolyze triacylglycerols in liver.
E. hydrolyze triacylglycerols in brain.

11. A readily-available intermediate in glycolysis is used to start the synthesis of the glycerol moiety of triacylglycerols. This intermediate is:

A. pyruvate.
B. 3-phosphoglyceric acid.
C. glycerol phosphate.
D. phosphoenolpyruvate.
E. dihydroxyacetone phosphate.

12. The lipoprotein that accumulates in blood when receptors on membranes of cells of peripheral tissues are defective, preventing endocytosis, is:

A. chylomicrons.
B. HDL
C. IDL.
D. LDL.
E. VLDL.

13. The complete oxidation of palmitic acid to carbon dioxide and water leads to the formation of a net total of high energy bonds of

A. 39 ATP
B. 69 ATP
C. 99 ATP
D. 129 ATP
E. 159 ATP

14. The complete hydrolysis of cardiolipin results in the release of

A. 0 mole of fatty acid.
B. 1 mole of fatty acid.
C. 2 moles of fatty acid.
D. 3 moles of fatty acid.
E. 4 moles of fatty acid.

15. The diet must provide

A. lecithin.
B. sphingomyelin.
C. cerebrosides.
D. linoleic acid.
E. phosphatidic acid.

16. Mitochondrial membranes are permeable to

A. fatty acyl ACP.
B. fatty acyl-CoA.
C. malonyl-CoA.
D. acetyl-CoA.
E. none of the above.

17. The sequence of repeated reactions in synthesis of fatty acids is best described by:

A. reduction - dehydration - oxidation - condensation.
B. dehydration - reduction - condensation - reduction.
C. condensation - reduction - dehydration - reduction.
D. condensation - dehydration - reduction - reduction.
E. condensation - reduction - oxidation - reduction.

18. All glycerol-containing lipids are synthesized from

A. triglyceride.
B. cephalin.
C. phosphatidic acid.
D. diglyceride.
E. monoglyceride.

19. Ketosis is the consequence of increased blood levels of

A. acetoacetate.
B. acetyl-CoA.
C. β-hydroxy-β-methyl-glutarate.
D. all of the above.
E. none of the above.

20. Essential fatty acids and precursors of prostaglandins include:

A. palmitic acid.
B. lignoceric acid.
C. linolenic acid.
D. oleic acid.
E. myristic acid.

21. All of the following statements are true EXCEPT:

A. High density lipoproteins (HDL) are involved in the transport of cholesterol from the periphery to the liver.
B. A high level of HDL-cholesterol is detrimental to one's health.
C. The major protein moieties of HDL are apolipoproteins A-I and A-II.
D. HDL appears to be antiatherogenic.
E. Chylomicrons are the largest and least dense of the lipoproteins.

22. Triacylglycerols of chylomicrons are:

A. intact triglycerides of intestinal lumen.
B. newly synthesized by intestinal cells.
C. transported from the liver.
D. converted to HDL.
E. none of the above.

23. An isomerase is needed for:

A. biosynthesis of saturated fatty acids.
B. biosynthesis of steroids.
C. oxidation of some unsaturated fatty acids.
D. oxidation of saturated fatty acids.
E. catabolism of steroids.

24. Substances requiring bile salt for transport into intestinal cells include:

A. vitamin A.
B. vitamin B.
C. vitamin C.
D. glucose.
E. amino acids.

25. In metabolism of fatty acids in adipose tissue,

A. insulin promotes fatty acid synthesis and triacylglycerol synthesis.
B. catecholamine promotes synthesis of triacylglycerol.
C. glucagon inhibits hormone-sensitive lipase.
D. insulin activates hormone-sensitive lipase.
E. catecholamine promotes synthesis of fatty acids.

26. β-Oxidation is generally referred to as the process of:

A. oxidation of the 3β-hydroxyl group of cholesterol to a keto-group.
B. oxidation of an alcohol to an acid.
C. oxidation using mono-oxygenase as catalyst.
D. all of the above.
E. none of the above.

27. During fatty acid synthesis, $^{14}CO_2$ is incorporated into the carboxyl of malonyl-CoA. At the condensation step, this $^{14}CO_2$ is:

A. incorporated into fatty acid.
B. incorporated into acetyl-CoA being synthesized.
C. eliminated as $^{14}CO_2$.
D. transferred to biotin.
E. none of the above.

28. The *de novo* biosynthesis of triacylglycerols occurs mainly in

A. liver and brain.
B. brain and muscle.
C. adipose tissue and muscle.
D. muscle and liver.
E. liver and adipose tissue.

29. The major energy source for the brain is normally

A. blood glucose.
B. blood amino acid.
C. blood ketone bodies.
D. blood fatty acids.
E. blood lactic acid.

30. Oxidation of fatty acids occurs in or on the:

A. mitochondrial matrix.
B. outer mitochondrial membrane.
C. cytosol.
D. intermembrane space.
E. Golgi apparatus.

31. In the fed state, when fatty acids are being synthesized in liver they cannot immediately undergo β-oxidation. This is because:

A. cAMP and fatty acyl CoA depress the activity of acetyl CoA carboxylase.
B. malonyl CoA inhibits carnitine-acyl transferase on the inner mitochondrial membrane.
C. fatty acyl CoA inhibits mitochondrial citrate synthase.
D. Insulin leads to increased uptake of glucose into cells and an increased rate of glycolysis.
E. gluconeogenesis takes precedence over fatty acid oxidation.

32. Gangliosides contain

A. neuraminic acid.
B. galactose.
C. fatty acid.
D. sphingosine.
E. all of the above.

33. Carnitine acyl transferase is necessary for:

A. fatty acid oxidation.
B. fatty acid synthesis.
C. cholesterol synthesis.
D. cholesterol oxidation.
E. digestion of triacylglycerols.

34. Which of the following is required for activation of diacylglycerol phosphate?

A. ATP
B. dATP
C. GTP
D. UTP
E. CTP

35. A fatty acid with an odd number of carbon atoms will enter the citric acid cycle in part as

A. citrate.
B. isocitrate.
C. α-ketoglutarate.
D. succinate.
E. malate.

36. Which of the following carries 2-carbon units from mitochondria to cytosol for fatty acid synthesis?

A. oxaloacetate.
B. malate.
C. pyruvate.
D. carnitine.
E. citrate.

37. The role of lipoprotein lipase is to:

A. digest dietary lipoproteins in intestinal lumen.
B. mobilize of dietary fat.
C. carry out intracellular lipolysis of lipoproteins.
D. hydrolyze triacylglycerols of plasma lipoproteins for transport of the released fatty acids into tissues.
E. none of the above.

38. Fatty acids in transport from adipose tissue to energy-utilizing tissues like muscle occur in the blood as:

A. chylomicrons.
B. free fatty acids bound to albumin.
C. VLDL.
D. HDL.
E. none of the above.

39. Mitochondria do NOT contain:

A. cytochrome oxidase
B. citrate synthase
C. HMG CoA synthase
D. β-ketoacyl CoA thiolase
E. acetyl CoA carboxylase

40. Membranes may contain all EXCEPT:

A. phosphatidyl choline
B. phosphatidyl ethanolamine
C. cerebroside (a glycolipid)
D. triacylglycerol
E. cholesterol

41. An intermediate in synthesis of BOTH triacyl-glycerols and phosphoglycerides is:

A. phosphatidic acid (phosphatidate).
B. UDP-glycerol.
C. CDP-diacylglycerol.
D. phosphatidyl serine.
E. sphingomyelin.

42. The rate-controlling step in the synthesis of fatty acids is catalyzed by:

A. acetyl CoA carboxylase
B. fatty acid synthase.
C. HMG CoA synthase.
D. hormone-sensitive lipase.
E. biotin.

43. Which of the following processes would be most active in the fasting state?

A. Activation of cAMP-dependent hormone-sensitive lipase in adipose tissue.
B. Lipolysis of triacylglycerols of chylomicrons by lipoprotein lipase in capillaries.
C. Activation of acetyl CoA carboxylase by citrate in liver.
D. Phosphorylation of glycerol by glycerol kinase in adipose tissue.
E. Production of NADPH by the hexose monophosphate pathway in adipose tissue.

44. Activation of fatty acids requires:

A. ATP
B. dATP
C. GTP
D. UTP
E. CTP

45. The immediate precursor of prostaglandins is:

A. myristate
B. palmitate
C. linoleate
D. linolenate
E. arachidonate

46. Aspirin inhibits the synthesis of

A. prostaglandins.
B. steroid hormones.
C. phosphatidic acid.
D. lipoproteins.
E. epinephrine.

47. Conversion of a diglyceride to lecithin requires:

A. UDP-glucose.
B. CDP-choline.
C. ACP-fatty acid ester.
D. malonyl-CoA.
E. none of the above.

48. Acetyl-CoA carboxylase

A. is activated by citrate.
B. is the rate limiting step in fatty acid synthesis.
C. contains biotin.
D. is inhibited by fatty acyl CoA.
E. all of the above.

49. Ketone bodies

A. can be a source of energy for muscle.
B. can be a source of energy for brain in starvation.
C. are made during periods of rapid lipolysis.
D. can cause a metabolic acidosis in diabetics.
E. all of the above are correct.

50. Hydrolysis of sphingomyelin would yield all of the following EXCEPT:

A. phosphate
B. choline
C. fatty acid
D. sphingosine
E. glycerol

51. Cholestyramine exerts its hypocholesterolemic effect by interaction with:

A. cholesterol.
B. triacylglycerols.
C. lipoproteins.
D. bile salts.
E. cholesterol esters.

52. The lipoprotein formed by intestinal cells from dietary lipid is:

A. chylomicrons.
B. HDL
C. IDL.
D. LDL.
E. VLDL.

53. The last step in a cycle of β-oxidation is catalyzed by:

A. 3-*cis* → 2-*trans* enoyl CoA isomerase
B. acyl CoA dehydrogenase.
C. acyl CoA thiolase.
D. hydroxymethyl glutaryl (HMG) CoA synthase.
E. hydroxyacyl CoA dehydrogenase.

54. Intermediates in the degradation of fatty acids are linked to:

A. acyl carrier protein.
B. coenzyme A.
C. CDP.
D. ATP.
E. pyrophosphate.

55. Hydrolysis of lecithin would yield all of the following EXCEPT:

A. phosphate
B. choline
C. fatty acid
D. sphingosine
E. glycerol

XII. ANSWERS TO QUESTIONS ON LIPIDS

1.	C	12.	D	23.	C	34.	E	45.	E
2.	B	13.	D	24.	A	35.	D	46.	A
3.	C	14.	E	25.	A	36.	E	47.	B
4.	E	15.	D	26.	E	37.	D	48.	E
5.	E	16.	E	27.	C	38.	B	49.	E
6.	B	17.	C	28.	E	39.	E	50.	E
7.	D	18.	C	29.	A	40.	D	51.	D
8.	B	19.	A	30.	A	41.	A	52.	A
9.	B	20.	C	31.	B	42.	A	53.	C
10.	C	21.	B	32.	E	43.	A	54.	B
11.	E	22.	B	33.	A	44.	A	55.	D

8. STEROIDS

Thomas Briggs

I. CHOLESTEROL: STRUCTURE AND CHEMISTRY

Carbon skeleton. The full system in cholesterol consists of four fused rings, two "angular methyl groups," and an 8-carbon side-chain, 27 carbons in all (Figure 8-1). In many steroid derivatives the side-chain is reduced or absent.

Substitution. Saturated carbons of the ring system may bear substituents which project on either side of the plane of the ring. In structural formulas, these are designated as follows:

> · · · (dotted line) = α (a̲lpha, a̲way from observer).

> —— (solid line) = β (b̲eta, toward, or b̲y observer).

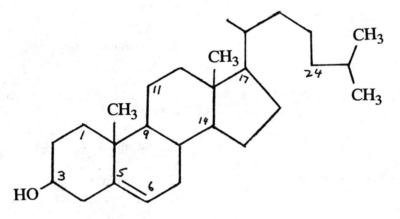

Figure 8-1. Cholesterol.

The **3β-hydroxyl** group, though hydrophilic, cannot overcome the non-polarity of the rest of the molecule, hence cholesterol is **insoluble in water.**

The $\Delta^{5\text{-}6}$ double bond can be reduced in two ways:

5α-H: 5α-cholestan-3β-ol; new hydrogen *trans* to the angular methyl group (C-19); rings A/B *trans.*

5β-H: 5β-cholestan-3β-ol; new hydrogen *cis* to C-19; rings A/B *cis.* This is mainly a bacterial product (coprosterol in older usage, from Gr. κοπροσ, meaning feces, where it is found). But the 5β-configuration occurs in some derivatives such as the bile acids and certain metabolites of steroid hormones.

II. OCCURRENCE AND FUNCTION

The name means **"bile-solid-alcohol."** It is found in bile, (a frequent component of gallstones, where it was first discovered), and is a crystalline solid and an alcohol.

Cholesterol occurs in **all cells** and tissues of higher organisms, but is especially abundant in nervous tissue and egg yolk. Related sterols are found in plants and higher microorganisms; these are poorly absorbed from the digestive tract. For storage, sterols are often **esterified** with unsaturated fatty acids. The total amount in a human averages about 180 grams. A typical value for blood cholesterol in developed countries is 180 mg/100 ml. Note that on a weight basis this is twice as high as the level of blood glucose.

Cholesterol has a universal function as a **component of membranes**, with a complex effect on fluidity (see chapter 9). Also it is a **precursor** of many important biological substances (see **bile acids, hormones**). In human disease, its insolubility makes deposits troublesome, especially in **atherosclerotic plaques** and **gallstones**.

In blood, cholesterol is carried as a complex with **lipoproteins**, partly esterified with a fatty acid, and partly with the 3-OH free. (See Chapter 7, section IX). The cholesterol in blood can exchange readily with liver cholesterol, but some cholesterol pools, especially that in brain, exchange very slowly.

III. BIOSYNTHESIS OF CHOLESTEROL

A. Acetate to Squalene

Formation of β-hydroxy-β-methyl-glutaryl CoA (HMG CoA)

    2 acetyl CoA $\rightleftharpoons$ acetoacetyl CoA + CoA (as in ketogenesis)

    acetoacetyl CoA + acetyl CoA $\rightleftharpoons$ β-hydroxy-β–methyl-glutaryl CoA, or HMG CoA

Reduction of HMG CoA to Mevalonic Acid

$$\text{HMG CoA} + 2\,\text{NADPH} + 2\,\text{H}^+ \xrightarrow[\text{Reductase}]{\text{HMG CoA}} \text{MVA} + 2\,\text{NADP}^+ + \text{HSCoA}$$

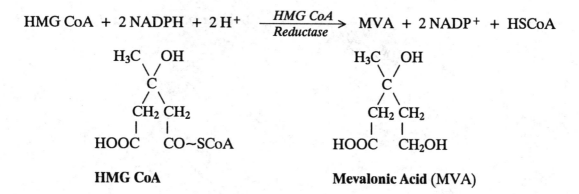

HMG CoA   **Mevalonic Acid (MVA)**

This reaction is the first committed, largely irreversible step in cholesterol biosynthesis, and is an **important control point**. The enzyme is rate-limiting for the entire pathway, and is inhibited by the eventual end-product, cholesterol.

Then mevalonic acid is converted to **isopentenyl pyrophosphate**, which is the "active isoprene unit": the precursor of all isoprenoid compounds such as terpenes; vitamins A, D, E, K; sterols; rubber.

$$\text{MVA} \xrightarrow[\text{3 times}]{\text{ATP, Mg}^{++}} \quad \begin{array}{c} \text{H}_2\text{C} = \text{C-CH}_2\text{-CH}_2\text{OPP} \\ | \\ \text{CH}_3 \end{array}$$

    Isopentenyl Pyrophosphate

    (CO_2 is lost)

Isopentenyl-PP can isomerize to **dimethylallyl-PP**, which condenses with another isopentenyl-PP to form **geranyl-PP**, a monoterpene (C-10):

$$\begin{array}{c} \text{H}_2\text{C} = \text{C-CH}_2\text{-CH}_2\text{OPP} \\ | \\ \text{CH}_3 \end{array} \rightleftharpoons \begin{array}{c} \text{H}_3\text{C-C} = \text{CH-CH}_2\text{OPP} \\ | \\ \text{CH}_3 \end{array} \longrightarrow \begin{array}{c} \text{H}_3\text{C-C} = \text{CH-CH}_2\text{CH}_2\text{C} = \text{CH-CH}_2\text{OPP} \\ | \hspace{3.5cm} | \\ \text{CH}_3 \hspace{3cm} \text{CH}_3 \end{array}$$

Geranyl-PP condenses with another *i*-pentenyl-PP to form **farnesyl-PP**, a sesquiterpene (C-15).

Two farnesyl-PP condense head-to-head (NADPH required) to form **squalene**, a triterpene (C-30).

B. Squalene to Cholesterol *(Figure 8-2)*

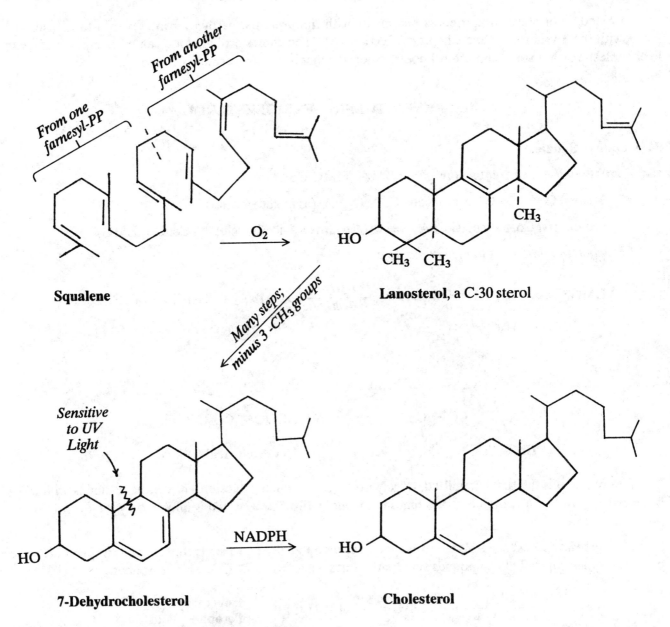

Figure 8-2. Conversion of **Squalene to Cholesterol**

Ring B of 7-dehydrocholesterol can be split by UV light (at arrow, Figure 8-2) to form vitamin D₃. The presence of 7-dehydrocholesterol in human skin explains how sunlight leads to the formation of vitamin D. In the body, vitamin D (cholecalciferol) is transported to the liver and hydroxylated at C-25, then to the kidney and hydroxylated at 1α, to form **1α,25-dihydroxycholecalciferol**, or **calcitriol**. *This is the biologically active form.* Vitamin D can be regarded as a hormone because it acts on target cells by the same mechanism as the steroid hormones. The active form promotes calcium absorption in the intestine by stimulating the synthesis of a **calcium binding protein**.

IV. METABOLISM OF CHOLESTEROL

A. Secretion

Some cholesterol is simply **secreted as is** into intestine. A little free cholesterol also occurs in bile.

B. Conversion to Bile Acids *(Figure 8-3)*

Quantitatively this is the major route for the excretion of cholesterol in humans.

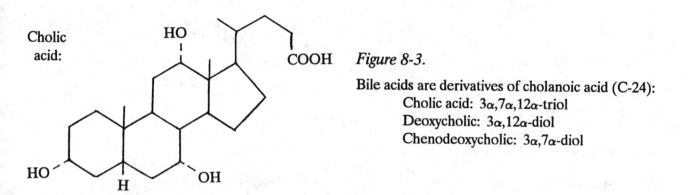

Cholic acid:

Figure 8-3.

Bile acids are derivatives of cholanoic acid (C-24):
Cholic acid: 3α,7α,12α-triol
Deoxycholic: 3α,12α-diol
Chenodeoxycholic: 3α,7α-diol

Bile acids are formed from cholesterol by the liver. In human bile they occur as conjugates of **glycine** or **taurine** (called bile salts). Together with a phospholipid such as lecithin (phosphatidyl choline), they aggregate as **mixed micelles** above a critical micellar concentration (CMC). These function as detergents and promote the emulsification, lipolysis, and absorption of fats, including the fat-soluble vitamins.

Bile salts are extensively reabsorbed (>95%) in the ileum, and *via* the portal system, undergo **entero-hepatic circulation** several times per day. Bacterial action in the gut may result in some structural changes. Insufficient bile salt secretion can be a cause of gallstone formation because the cholesterol that normally occurs in bile needs bile salts to stay in solution. Continued disturbance of BS metabolism can lead to malabsorption syndromes and, in extreme cases, deficiency of fat-soluble vitamins. Oral chenodeoxycholic acid has been useful as replacement therapy to supplement the bile acid pool (well in excess of the CMC) to the point where cholesterol gallstones may redissolve.

C. Steroid Hormones

Production of hormones from cholesterol is quantitatively minor, but of major importance physiologically.

Major types: The examples shown in Figure 8-4 (next page) are the principal members of each major category (gonadal and adrenal) secreted in the human.

Formation from Cholesterol: All steroid-producing tissues cleave the side-chain of cholesterol between carbons 20 and 22 to form **pregnenolone** (enzyme: 20,22-lyase, or desmolase). This is the rate-limiting step. In most cases, pregnenolone is then converted to **progesterone**.

    a. Gonads: The *corpus luteum* stops at progesterone. The testis, after 17-hydroxylation, cleaves the remaining side-chain (enzyme: 17,20-lyase) to form **19-carbon** steroids (androstenedione, then testosterone). The ovary also does this, in fact makes testosterone, but then aromatizes ring A (resulting in the loss of carbon 19) to form **18-carbon** steroids (estradiol, estrone).

Gonadal steroids

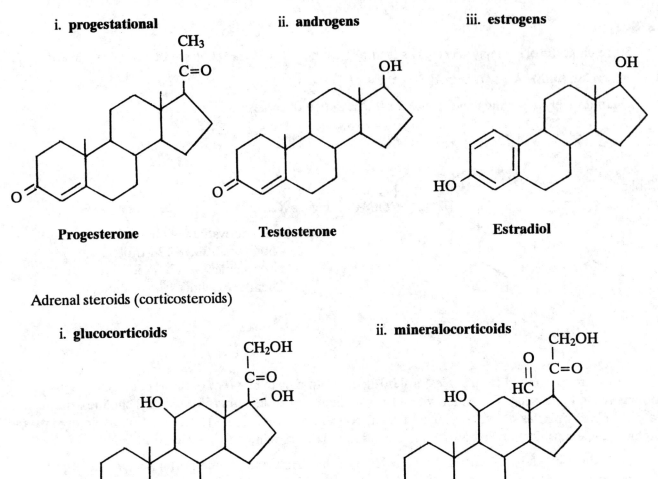

Figure 8-4. Representative Steroid Hormones in the Human

b. Adrenal cortex: hydroxylations occur in sequence at positions **17, 21, 11, 18**. Except: the *zona fasciculata*, which makes cortisol, has no 18-hydroxylase, and the *zona glomerulosa*, which makes aldosterone, has no 17-hydroxylase. Some conversion to androgens (especially **dehydroepiandrosterone**, or **DHEA**) and estrogens may also occur. In a patient with Cushing's syndrome, these androgens may cause signs of virilization such as hirsutism.

Metabolism: Reductions occur involving ring A, and the products are excreted in bile and/or urine as conjugates of **sulfate** or **glucuronic acid**. Androgens are oxidized at C-17 to ketones of the class known as **17-ketosteroids**. Urinary 17-KS are indicative of androgen metabolism of *both* testicular and adrenal origin.

V. ACTION OF STEROID HORMONES

A. Mechanism

In contrast to most peptide hormones, which interact with a receptor on the plasma membrane without penetrating the cell, steroid hormones **enter target cells** and are bound by a **receptor** in the **cytoplasm**. The hormone-receptor complex then diffuses to the nucleus and binds to chromatin of selected genes, inducing (or repressing) the **expression** of particular **genes**.

B. Function

Gonadal Hormones

a. **testosterone** promotes development of male secondary sex characteristics, and also has anabolic activities. With FSH, it promotes spermatogenesis by seminiferous tubules.

b. **estradiol** promotes development of female secondary sex characteristics, proliferative phase of endometrium, and, with FSH, development of ovarian follicle and finally, ovulation.

c. **progesterone** supports secretory phase of endometrium, luteal phase of ovary, and inhibits further ovulation.

Adrenal hormones

a. **Glucocorticoids** promote gluconeogenesis by inducing key enzymes such as pyruvate carboxylase. Proteins, especially those of muscle, are broken down to amino acids, which the liver converts into glucose which is partly released to the circulation and partly stored as liver glycogen.

Cortisol (also known as hydrocortisone) and synthetic steroids such as prednisone and prednisolone also have anti-inflammatory and anti-immune effects. When used in high doses over a long time they cause, besides muscle wasting, lipolysis on the extremities but accumulation of fat on the face and trunk ("Cushingoid" features).

b. **Mineralocorticoids** promote retention of Na^+, along with H_2O, by the kidney, and excretion of H^+ and K^+. A high level of **aldosterone** causes hypertension; deficiency, excessive loss of salt.

C. Regulation

Gonadal hormones: Trophic hormones are the same in both sexes. The hypothalamic **Gonadotropin Releasing Hormone (GnRH)**, which is the same as **Luteinizing Hormone Releasing Hormone (LHRH)**, stimulates the pituitary to release both **Follicle Stimulating Hormone (FSH, formerly known as ICSH)** and **Luteinizing Hormone (LH)**.

In the male, FSH induces spermatogenesis in seminiferous tubules; feedback is by a glycoprotein, **inhibin**. LH stimulates production of **testosterone** by Leydig cells. Feedback is on both pituitary and hypothalamus.

In the female, FSH induces follicular development and production of **estradiol**. At mid-cycle, there is a positive feedback effect by estrogen, causing a surge of both FSH and LH production by the pituitary. Luteinization ensues, with production of **progesterone** by the *corpus luteum*.

Adrenal hormones

a. The hypothalamus puts out **Corticotropin Releasing Hormone (CRH)** which stimulates the pituitary to release **Adrenocorticotropic Hormone (ACTH)**, which then signals the adrenal cortex to secrete **cortisol**. Negative feedback by cortisol occurs on both pituitary and hypothalamus. In extreme cases, such as when steroid drugs are used in high doses for a long time, adrenal **atrophy** may occur due to prolonged inhibition of ACTH production. Conversely, lack of cortisol production, as in an enzymatic defect in the adrenal, may lead to adrenal **hyperplasia** due to excessive ACTH production in an attempt by the pituitary to compensate for the lack of steroid hormone.

        b. Production of **aldosterone** is regulated largely by the renin-angiotensin system. In response to a perceived drop in perfusion pressure, the kidney produces **renin**, which converts **angiotensinogen**, a peptide made by the liver, to angiotensin I. Then this is converted by an enzyme in lung tissue to **angiotensin II**, which induces production of aldosterone in the *zona glomerulosa* of the adrenal. Potassium also induces formation of aldosterone. Thus an excess of renin or aldosterone causes hypertension; deficiency results in salt loss.

VI. CHOLESTEROL LEVELS

Because of its insolubility in an aqueous medium, cholesterol must be "packaged" as part of a lipoprotein in order to be transported in blood (see chapter on Lipids). But since a high level of blood cholesterol, especially that contained in LDL, is a risk factor in atherosclerotic heart disease, much attention is being given to factors that lower cholesterol levels:

Diet: Cholesterol itself can be avoided by substituting vegetable products for meat and dairy products. In addition, consuming more polyunsaturated fats and fewer saturated fats has a cholesterol-lowering effect. Fats containing monounsaturates seem to be benign. There are exceptions to these generalizations.

Exercise: This activity seems to increase the HDL/LDL ratio, which correlates negatively with atherosclerosis.

Drugs: Some that have been helpful are: **nicotinic acid** and fibric acid derivatives such as **clofibrate**, which lower cholesterol levels by poorly-understood mechanisms; **compactin**, **mevinolin**, and **lovastatin**, which inhibit HMG CoA reductase, the controlling step in cholesterol synthesis; and **cholestyramine**, a resin which, taken orally, binds bile acids and promotes their excretion rather than enterohepatic circulation, causing the liver to replace them by new synthesis from cholesterol, some of which comes from the blood.

VII. DISORDERS OF CHOLESTEROL AND STEROID METABOLISM

A. Familial Hypercholesterolemias

This group of conditions results from a heritable lack of functional receptors for LDL. Since endocytosis of LDL cannot occur, intracellular HMG CoA reductase becomes more active, while at the same time blood levels of LDL reach very high levels. (These are discussed more fully in the chapter on Lipids). Diet and drugs may help the individual heterozygous for the condition; little can be done for the homozygote.

B. Inborn Errors of Steroid Metabolism

Female Pseudohermaphroditism (adrenogenital syndrome)

    a. *21-Hydroxylase* deficiency

        Molecular defect: Deficient 21-Hydroxylase; several mutant types.
        Pathway affected: C-21 hydroxylation.
        Diagnosis: Increased 17-ketosteroids and 21-deoxysteroids (blood and urine).
        Genetics: Autosomal recessive.

b. Other enzyme defects causing the adrenogenital syndrome

Deficient Enzyme	Compound	Genetics
11β-Hydroxylase	17-Ketosteroids (urine)	Autosomal recessive.
3β-Hydroxysteroid Dehydrogenase	DHEA (blood)	Autosomal recessive.

Male Pseudohermaphroditism, androgen deficiency

a. *5α-Reductase* deficiency

Molecular defect: Deficient enzyme from more than one mutation.
Pathway affected: Testosterone → Dihydrotestosterone (DHT).
Diagnosis: Elevated ratio of testosterone/DHT.
Genetics: Autosomal recessive.

b. Other enzyme defects

Deficient Enzyme	Compound	Genetics
20,22-Desmolase	All steroids	Autosomal recessive
17-Hydroxylase	Testosterone low	Autosomal recessive
17,20-Desmolase	Pregnanetriolone	
17-Dehydrogenase	Δ^4-androstenedione / testosterone ratio	

c. Androgen receptor defects: complete and incomplete testicular feminization; Reifenstein syndrome and the infertile male syndrome.

Molecular defect: Abnormal or absent androgen receptor protein.
Pathway affected: Binding of androgens to receptors.
Diagnosis: Absent binding of DHT to receptors.
Genetics: X-linked recessive.

VIII. CHOLESTEROL IN PERSPECTIVE

Cholesterol has many roles in health and disease, both in its own right, and as a precursor of a variety of biologically-important substances (Figure 8 - 5):

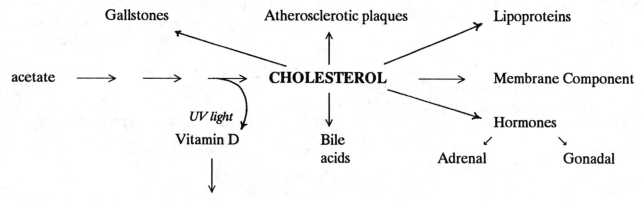

Figure 8-5. The many functions of cholesterol.

IX. REVIEW QUESTIONS ON STEROIDS

DIRECTIONS: For each of the following multiple-choice questions (1 - 21), choose the ONE BEST answer.

1. In humans,

A. all plasma cholesterol is carried as cholesterol esters.
B. cholesterol is made only by liver cells.
C. cholesterol in brain exchanges readily with cholesterol in plasma.
D. cholesterol is entirely metabolized to CO_2 and H_2O.
E. conversion of cholesterol to bile acids involves shortening the side chain.

2. Which of the following stimulates the synthesis of calcium binding protein in intestinal mucosa?

A. calcium
B. 1,25-dihydroxycholecalciferol
C. alpha-tocopherol
D. calcitonin
E. parathyroid hormone.

3. Estriol is excreted in the urine as the conjugate with

A. glucuronic acid
B. cysteine
C. glycine
D. glutamine
E. protein.

4. In fasting there is lowered activity of liver HMG CoA reductase and squalene oxidocyclase, and therefore less synthesis of

A. HMG CoA
B. ubiquinone
C. acetoacetate
D. retinol
E. cholesterol.

5. In mammals, which of the following can NOT take place?

A. estrone → estradiol
B. estrone → dihydrotestosterone
C. 17-hydroxypregnenolone → testosterone
D. progesterone → estrogen
E. cholesterol → estrogen.

6. Serum cholesterol levels may be significantly changed by a diet with a low ratio of:

A. diglycerides to monoglycerides
B. saturated to polyunsaturated fatty acids
C. lactose to glucose
D. vitamin K to vitamin D
E. cephalin to lecithin.

7. The surge of gonadotropins around the time of ovulation

A. is caused by the effect of GnRH on the ovary
B. is caused in part by a negative feedback, acting on the hypothalamus, of estradiol secreted during the luteal phase
C. starts the development of follicles that release estradiol during the luteal phase
D. starts the development of the *corpus luteum* that releases progesterone during the luteal phase.
E. all of the above are correct.

8. Calcitriol is the most active form of

A. vitamin K
B. vitamin E
C. vitamin D
D. vitamin B_{12}
E. vitamin A.

9. Some virilization may be seen in a patient with Cushing's syndrome, but not in a patient on high doses of prednisone because

A. prednisone has weak glucocorticoid activity and prolonged use does not cause adrenal atrophy
B. prednisone has weak mineralocorticoid activity and is a strong glucocorticoid
C. prednisone cannot produce a Cushingoid toxicity syndrome
D. virilization in the patient with Cushing's syndrome is caused by adrenal androgens
E. Cushing's syndrome is an autosomal dominant trait.

10. In man, cholesterol

A. is converted to glycocholic acid by intestinal bacteria.
B. is a part of cell membranes
C. can be catabolized mainly to acetyl CoA.
D. is a precursor of squalene
E. is excreted mainly as the glucuronide.

11. All of the following statements concerning Vitamin D are true EXCEPT:

A. It is not itself active in stimulating calcium transport by intestine and calcium mobilization from bone *in vivo*.
B. It can be produced by ultraviolet light acting on $\Delta^{5,7}$ sterols.
C. It is converted to the 1,25-dihydroxy-derivative which stimulates the intestinal mucosa to transport calcium.
D. It induces the kidney to excrete phosphate.
E. It is acted upon by liver and then by kidney.

12. A lack of receptors for LDL will probably induce

A. a low level of blood LDL
B. a low activity of HMG CoA reductase
C. decreased plasma cholesterol
D. decreased endocytosis of LDL
E. increased activity of intracellular ACAT (cholesterol esterifying enzyme).

13. Which of the following promotes mineralization of bone?

A. decreased levels of ascorbic acid
B. decreased levels of plasma calcium
C. parathyroid hormone
D. 1,25-dihydroxycholecalciferol
E. sodium taurocholate.

14. High doses of hydrocortisone over 2 months would probably cause

A. a decrease in liver glycogen
B. larger skeletal muscles
C. atrophy of the adrenal cortex
D. a rise in blood ACTH
E. decreased gluconeogenesis.

15. All of the following depend on micellar activity for their absorption EXCEPT

A. glycine
B. vitamin E
C. cholesterol
D. vitamin A
E. stearic acid.

16. The mechanism of action of progesterone involves all of the following EXCEPT:

A. interaction with a cytoplasmic receptor
B. penetration of the hormone into the cell
C. activation of adenylate cyclase
D. synthesis of new protein
E. altered genetic expression.

17. Cholestyramine lowers blood cholesterol levels because:

A. it interrupts the efficient enterohepatic circulation of bile salts.
B. it inhibits HMG CoA reductase.
C. it depresses triglyceride synthesis, thereby lowering levels of lipoproteins in blood.
D. it induces the production of receptors for LDL.
E. it depresses the production of VLDL by liver.

18. Biosynthesis of cholesterol involves all of the following EXCEPT:

A. dimethylallyl pyrophosphate
B. succinyl CoA
C. isopentenyl pyrophosphate
D. squalene
E. lanosterol.

19. Cholesterol can be converted to

A. androgens
B. corticosteroids
C. bile acids
D. glucocorticoids
E. all of the above.

20. The biosynthesis of testosterone involves all of the following EXCEPT:

A. isopentenyl pyrophosphate
B. pregnenolone
C. estradiol
D. lanosterol
E. cholesterol.

21. Agents that lower blood cholesterol include

A. cholestyramine
B. mevinolin
C. clofibrate
D. nicotinic acid
E. all of the above.

MATCHING: For each set of questions, choose the ONE BEST answer from the list of lettered options above it. An answer may be used one or more times, or not at all.

Questions 22 - 25:

A. Lanosterol
B. 7-Dehydrocholesterol
C. Calcitriol
D. Sodium taurocholate
E. Pregnenolone
F. Progesterone

G. Testosterone
H. Dihydrotestosterone
I. Estradiol
J. Cortisol
K. Aldosterone
L. Dehydroepiandrosterone

22. Androgen produced by action of 5α-reductase in some target tissues.

23. Most potent glucocorticoid in humans.

24. Deficient in a patient with rickets.

25. Precursor of elevated urinary 17-ketosteroids in a patient with an adrenal tumor.

X. ANSWERS TO QUESTIONS ON STEROIDS

1. E
2. B
3. A
4. E
5. B
6. B
7. D
8. C
9. D
10. B
11. D
12. D
13. D

14. C
15. A
16. C
17. A
18. B
19. E
20. C
21. E
22. H
23. J
24. C
25. L

9. MEMBRANES
Thomas Briggs

I. OVERVIEW

A. What Membranes Are

A membrane is a **non-covalent assembly** of lipid and protein; often these major components also have carbohydrate residues attached to them. It is arranged as a **bilayered sheet**, with each layer being termed a **"leaflet."** Because interactions among components are non-covalent, each leaflet is **fluid** in that individual molecules are free to move within the plane of that leaflet. A membrane also is **asymmetric** in that one leaflet is different from the other, and many **directional functions** are carried out by the membrane.

B. What Membranes Do

Permeability barriers. A membrane forms the boundary between a cell and its environment, and also forms compartments within a cell. But a membrane is highly **selective** in what it allows to pass through it. Some molecules cross a membrane easily; other hardly at all, or only with the aid of a transport mechanism. The presence of transport mechanisms enables a membrane to selectively regulate what ions and molecules may pass through it.

Create and maintain gradients. Membranes not only regulate what may diffuse through, but also contain energy-dependent devices for creation of specific **gradients**, and for maintenance of different concentrations of a substance on each side.

Regulate flow of information. A membrane can have **receptors** on one side which are specific for a particular informational entity such as a hormone. Binding of the hormone leads to **transmission of a signal** across the membrane, with diverse physiological effects.

Convert energy. A membrane often has structural elements which convert one form of chemical energy to another. Examples: use of ATP energy to produce a gradient (as in **active transport**); use of a gradient to generate ATP (as in **oxidative phosphorylation**).

II. CHEMISTRY AND STRUCTURE

A. Lipid Components

Lipids of membranes are **amphipathic**, that is, they have both hydrophobic and hydrophilic characteristics. A typical amphipathic lipid has a polar "head group" containing a charge or other polar group, and a non-polar "tail" such as a long fatty acid chain or other hydrocarbon entity. The polar head-groups are attracted to water molecules, while the non-polar portions are repelled by water, and instead experience many weak interactions among themselves. The result (see below) can be a closed spherical structure such as a **micelle**, with hydrophobic portions sequestered in the interior, or a **bilayered sheet**, as occurs in membranes, with non-polar portions facing each other in the interior, and polar portions facing water on either side.

The total of many weak, non-covalent interactions confers sufficient stability on membrane-like structures that they can even self-assemble from constituent molecules. In this case, however, the asymmetry and directionality of natural membranes are lacking.

There are several types of amphipathic lipids:

1. *Phospholipids.* Most phospholipids are **phosphoglycerides**, ie., they are derivatives of phosphatidate (see Chapter 7). Examples are **phosphatidyl choline** (lecithin), and the phosphatidyl derivatives of glycerol, ethanolamine, and inositol. Diphosphatidyl glycerol (cardiolipin) is prominent in mitochondrial membranes.

One phospholipid that is not a derivative of phosphatidate is the sphingolipid, **sphingomyelin**. It is structurally very similar to lecithin, but consists of phosphoryl choline attached to ceramide, and thus has no glycerol.

2. *Glycolipids.* Other sphingolipids have carbohydrate residues as their polar head-groups. These may be one or several glucose, galactose, etc. units attached to ceramide, and are termed **cerebrosides**. If one or more of the sugars is an acid sugar (N-acetyl-neuraminic acid, or NANA) the lipid is a **ganglioside**.

3. *Cholesterol.* Included in spaces between hydrophobic fatty acid chains, cholesterol is oriented with the -OH group facing the water phase. See below (Section C-2) for its effect on a membrane's properties.

B. Proteins

Membranes contain 25 - 75% protein. **Intrinsic proteins** have much interaction with the membrane. They are stably embedded in the basic lipid bilayer and can only be removed with difficulty and by disrupting the integrity of the membrane. **Peripheral proteins** are weakly attached to the membrane, or to other proteins, and can be readily removed. Proteins are responsible for most of a membrane's functions.

Proteins of membranes typically have **domains** enriched in hydrophobic or in hydrophilic amino acid residues; these determine how the protein interacts with the membrane. For instance, an amino acid sequence with much leucine, valine, etc. is apt to coil into an α-helix which then becomes a membrane-spanning region of the protein. The non-polar R groups interact stably with the non-polar interior of the membrane. Amino acid sequences with many charged and polar R groups are likely to be part of those domains that are located in the aqueous medium on either side of the bilayer.

Membrane proteins may be **glycosylated**. The carbohydrate residues are often N-acetyl derivatives of **glucose** or **galactose**, attached to the N of asparagine or to the O of serine or threonine. Glycoproteins are usually located on the *extracellular* side of the bilayer.

C. Fluid Mosaic Model of Membrane Structure

Figure 9-1 summarizes current views regarding membrane structure. Amphipathic lipids are arranged in a bilayer, in which proteins are embedded. Major functional attributes of this structure are as follows:

1. *Fluidity.* The presence of *cis*-double bonds in naturally-occurring fatty acids is of critical importance. By introducing a bend in the hydrocarbon chain, a double bond prevents tight packing of chains, thus **lowering the melting point** of the lipid. Thus above a critical melting temperature (T_m) the fatty acid chains tend to be disordered, loosely packed, more mobile. This fluidity is favored by shorter chain lengths and by increased numbers of double bonds, whereas long saturated chains promote greater rigidity.

2. *Role of cholesterol.* The effect of cholesterol is complex. Inserted into spaces between fatty acid chains, it increases the density of packing, stiffening the membrane and reducing fluidity. But by preventing exact alignment or crystallization of lipid molecules, it prevents congealing of the membrane at low temperatures, and so it increases fluidity under some conditions. The effect is to **broaden T_m**.

3. *Mobility of components.* Individual lipid molecules are free to **diffuse laterally** within the plane of each leaflet. They are NOT free to cross ("flip-flop") to the opposite leaflet. The proteins, like icebergs in a sea, can also move laterally, but more slowly.

4. *Asymmetry.* As stated above, the two leaflets of a membrane are different. Generally, **glycolipids** are located on the **extracellular** side, as are **carbohydrate residues** on proteins. Directional functions, such as ion pumps, are inserted with proper directionality.

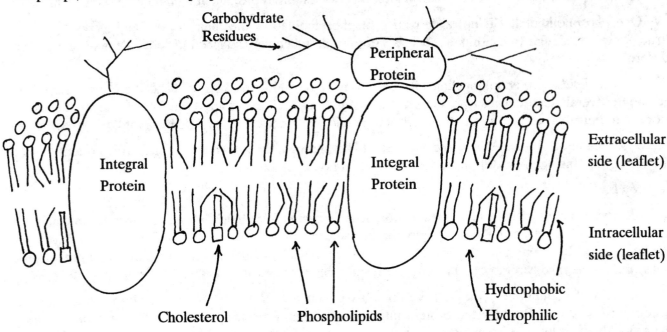

Figure 9-1. Fluid Mosaic Model of Biological Membranes.

III. FUNCTIONS

A. Barrier Function

With respect to diffusion, membranes are **impermeable to ions** and most polar molecules. Exceptions are H_2O and urea, which can cross a membrane freely, as do small gas molecules such as O_2, N_2, and CO_2. Membranes are readily **permeable to lipid-soluble substances** such as steroid hormones.

However, as described below, membranes contain a variety of highly selective transmembrane processes.

B. Transport Processes — Small Molecules

1. *Some definitions* (these concepts will be expanded and illustrated in subsequent sections):

a. **Passive diffusion:** movement of a substance down its electrochemical gradient in response to thermal motion of molecules, but limited by its solubility in the hydrophobic interior of the membrane (as measured by the permeability coefficient).

b. **Facilitated diffusion:** movement down an electrochemical gradient with the aid of a transport protein which may be specific and saturable. The rate may be very many times greater than that of passive (or simple) diffusion. No extra energy is required.

c. **Active transport:** movement of a substance against its electrochemical gradient (ΔG with respect to the substance is positive). A transport protein *and* input of energy are required.

d. **Uniport:** a system which moves only one type of molecule; movement may be bidirectional.

e. **Symport:** a system that moves two types of molecule in the same direction.

f. **Antiport:** a system that moves two types of molecule in opposite directions.

2. *Passive transport (simple diffusion).* In simple diffusion, molecules move from a region of **higher to** one of **lower concentration**, without involvement of any special transporting mechanisms. But the *rate* of diffusion may be severely limited by the solubility of a substance in a hydrophobic medium. Since the non-polar interior of the membrane is a hostile environment for a polar molecule to traverse, diffusion is very slow for many substances. This is illustrated by the extremely small permeability coefficients for glucose and chloride in simple artificial lipid bilayers (Table 9-1).

3. *Facilitated diffusion systems*

　　　a. **Ion pores** or **channels**. These provide a polar path through the membrane, so that hydrophilic substances are insulated from contact with the non-polar interior of the membrane. Gramicidin A is an example. Ions move down their electrochemical gradients, and no input of energy is required.

　　　b. **Transport proteins (permeases)**. These may be highly specific, such as the permease for glucose of erythrocyte membrane (Table 9-1). D-Glucose moves down its electrochemical gradient at an accelerated rate, and again, no input of energy is involved. L-Glucose hardly passes at all. This is an example of a **uniport**.

Other transport proteins may be **antiports**. An example is the anion exchange protein of erythrocyte membrane, which promotes the electrically neutral exchange of one ion (Cl^-) for another (HCO_3^-). The permeability coefficient for Cl^- is increased by a factor of 10,000,000 (Table 9-1).

Transported Species	Medium	Permeability Coefficient
Cl^-	Simple Lipid Bilayer	10^{-11}
Cl^-	Red Cell Membrane	10^{-4}
Glucose	Simple Lipid Bilayer	10^{-10}
Glucose	Red Cell Membrane	10^{-5}

Table 9-1. **Facilitated Diffusion** allows much faster penetration of red cell membrane by selected ions and small molecules.

　　　c. **Ionophores.** Some transport substances have a hydrophobic exterior with a hydrophilic cavity that may be quite specific for particular ions. An example is valinomycin, an antibiotic which specifically complexes and transports K^+ ions. Formerly it was thought that this complex would actually cross the membrane, taking the ion from one side to the other like a ferry boat. But this carrier model, demonstrated in artificial bilayers, may not be applicable in natural membranes.

4. *Characteristics of simple* vs. *facilitated diffusion (Figure 9-2).* Facilitated transport, being a process in which a protein is used to lower an activation-energy barrier, is not unlike an enzyme-catalyzed reaction. The protein may exhibit great specificity for the target molecule, and the rate is greatly increased compared with that of the unassisted process, especially at low concentrations of substrate. At higher concentrations the carrier, like an enzyme, can become saturated, and the rate of transport levels off. The kinetics, like enzyme kinetics, may be described in Michaelis-Menten terms, with K_M, and may exhibit competitive inhibition. These considerations do not apply to simple diffusion.

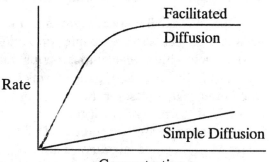

Figure 9-2. Effect of Concentration on Rate of Diffusion. Note saturation of facilitated system at higher concentrations of transported species.

5. *Active transport*. Since a substance is being forced to move against its electrochemical gradient, a source of energy is required. The overall ΔG is then negative.

a. **ATPases** use energy directly from the hydrolysis of ATP. A prime example is the **Na^+ - K^+ pump**. This is a Mg^{++}- requiring ATPase which has membrane-spanning subunits that undergo conformational changes in a directional manner so that for each ATP used, 3 Na^+ are extruded from the cell and 2 K^+ are brought in. It is inhibited by cardiotonic steroids such as ouabain and digitoxigenin.

Another important ATPase is the **calcium ion pump**, especially abundant in muscle tissue. Structurally and functionally related to the Na^+ - K^+ pump, it uses one ATP to transport two Ca^{++}. In muscle, it pumps Ca^{++} from cytosol into the sarcoplasmic reticulum. The **H^+ - K^+ ATPase** of stomach is also related to the sodium and calcium pumps.

The preceding are examples of **P-type ATPases**, so called because their mechanism of action involves a phosphorylated intermediate. Two other types occur: **V-type ATPases**, which transport H^+ and occur in membranes of lysosomes and of vesicles involved in endocytosis and exocytosis, and **F-type ATPases**, such as the F_o-F_1 ATPase of oxidative phosphorylation, which uses a proton gradient to *synthesize* ATP (see Chapter 4).

b. **Co-transport systems** can be driven by the energy of the Na^+ gradient, and thus only indirectly by ATP hydrolysis. Absorption of glucose from small intestine works in this way. A **Na^+ - glucose symport** protein is located in membranes of the apical surface of brush border cells, and obligatorily cotransports one glucose along with one Na^+. The Na^+ - K^+ ATPase is located only on the basolateral (serosal) membranes, and keeps the intracellular $[Na^+]$ very low, ensuring a concentration gradient between lumen and cell. Sodium ion, moving down its gradient, thus "drags" glucose with it into cells where the concentration of the latter may actually be high. Glucose then will pass into the bloodstream by facilitated diffusion.

Many other cells have similar uptake systems, especially for **amino acids**.

The sodium gradient is also linked to the transport of **calcium**, but in this case through an **antiport**, which pumps calcium out of cells, in exchange for 3 Na^+ entering down their gradient. This is particularly significant in heart muscle cells. A **Na^+ - H^+ antiport** is also used to regulate intracellular pH.

c. **Transport by modification** occurs when a substance such as a sugar is phosphorylated the instant it enters a cell. Though ΔG for initial entry may be unfavorable, phosphorylation effectively removes the product of the equilibrium (lowers the intracellular concentration of the free form), promoting entry of more sugar molecules since the overall ΔG is now negative (favorable) and the phosphorylated molecules are unable to escape.

C. Transport Processes — Large Molecules

Receptor-mediated endocytosis is a process well-illustrated by the uptake of LDL particles into extrahepatic cells (see also the chapter on Lipids). Small regions of the membrane become coated on the cytoplasmic side with a filamentous peripheral protein, **clathrin**, which participates in the formation of pits (concave to the outside of the cell) which are incipient vesicles. On the extracellular side, receptors congregate in the **coated pits**; for the uptake of LDL-cholesterol, these receptors are specific for apo B-100. Binding of LDL to receptors triggers internalization of the pits and bound LDL, which now become vesicles. Internalization is also triggered by Ca^{++}. The vesicles fuse with lysosomes, which release hydrolytic enzymes which degrade large molecules to smaller units for metabolism by the cell. Further endocytosis is regulated by a feedback mechanism which controls the number of receptors.

Membranes also engage in **exocytosis**, which in many ways is a reverse of endocytosis, and in **pinocytosis**, which results in entry of fluid and small solutes into the cell.

D. Transmembrane Signaling: Concepts

1. *Signal transduction by a "second messenger"* can be thought of as occurring in several distinct phases, as described in general terms below. This mechanism is applicable to the mode of action of a variety of agents, especially the **peptide hormones.** Specific illustrations will follow.

    a. The original stimulating agent, glucagon for example, binds to a **receptor** which is specific for it and is located only on the **extracellular side** of the membrane. The hormone remains where it is and *does not enter the cell.* But the receptor is a transmembrane protein, and binding of the hormone induces a conformational change which transmits a signal across the membrane to the cytoplasmic side.

    b. The transmitted conformational signal then activates an intermediary protein, one of the **"G proteins"** (see below).

    c. The activated intermediary protein then goes on to activate a special **enzyme.**

    d. The newly-activated enzyme produces a new intracellular entity, known as the **"second messenger,"** which is, in a sense, an intracellular hormone.

    e. The new entity proceeds to induce various intracellular effects, such as **phosphorylation** of certain target proteins, which, in turn, **activate or inhibit** various metabolic processes.

    f. Finally, there must be a way to reverse the procedure, **to turn the signal off.** A permanent "on" state, induced by some agents such as cholera toxin, leads to disease.

2. *G-Proteins* are the intermediary proteins that carry a signal from the intracellular side of the activated hormone receptor to the enzyme that catalyzes the formation of the second messenger. They bind guanyl nucleotides (hence the name), the complex with GDP being the inactive form, and consist of dissociable α- and other subunits. They comprise a family of related proteins, which are designated G_s if the are stimulatory to adenylate cyclase, or G_i if they inhibit adenylate cyclase. Other G-proteins work through different signaling systems.

On receipt of a hormonal signal, GTP replaces the bound GDP, causing the α-subunit to dissociate. The GTP-α-subunit is the active form that then activates (or inhibits) the enzyme that produces the second messenger. But the GTP-α-subunit has a "built-in" slow GTPase activity, which causes the bound GTP to revert to GDP, allowing re-association of subunits and inactivation of the signal. The process can be considered cyclic (Figure 9-3).

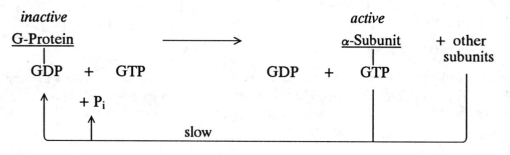

Figure 9-3. Dissociation - Reassociation Cycle of a G-Protein.

E. Transmembrane Signaling: Examples

1. *The Adenylate Cyclase System* is used with many peptide hormones. Some hormones that stimulate adenylate cyclase are ACTH, ADH, β-adrenergics, glucagon, PTH, etc. The list of agents that inhibit adenylate cyclase is shorter but includes acetylcholine, angiotensin II, somatostatin, and others. Notably, insulin does NOT involve this system.

Components of the system are (a) three proteins: the hormone **receptor** which is a transmembrane protein, a G_s (or G_i) **protein**, and **adenylate cyclase**; and (b) the second messenger itself, **cyclic AMP** (cAMP, shown in Figure 9-4, right).

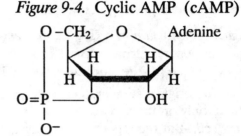

Figure 9-4. Cyclic AMP (cAMP)

Operation of the cAMP system is shown in Figure 9-5. Activation begins when a hormone (eg., glucagon) binds to the receptor on the outside of the cell. The signal is transmitted to the G-protein on the cytoplasmic side, which becomes activated and exchanges GTP for the bound GDP. Dissociation occurs, and the **GTP-α-subunit** then activates adenylate cyclase which promotes the reaction:

$$\text{ATP} \longrightarrow \text{cAMP} + \text{PP}_i$$

The cAMP then activates a **protein kinase** such as *phosphorylase kinase kinase* in muscle. Typically these cAMP-dependent kinases consist of **regulatory subunits** bound to **catalytic subunits**, the whole being inactive. Binding of cAMP to regulatory subunits causes them to dissociate, allowing the catalytic subunits to become active. These, in turn, go on to phosphorylate proteins, either activating or inactivating these, depending on whether the phosphorylated target protein is the catalytically active or inactive form. Since a variety of proteins may become phosphorylated, metabolic effects from a single hormonal stimulus may be quite diverse.

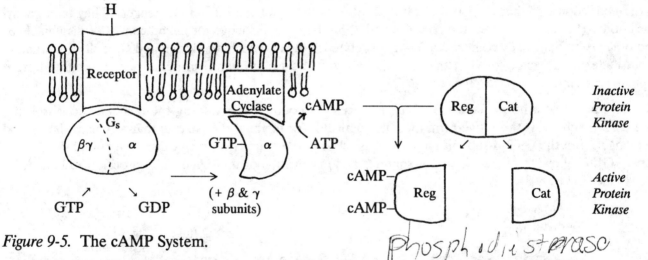

Figure 9-5. The cAMP System.

Inactivation of the system may occur in various ways: (a) a **phosphodiesterase** is present that degrades cAMP to AMP; (b) the G-protein slowly inactivates itself through its built-in **GTPase activity**; (c) **phosphatases** dephosphorylate the protein phosphates. All this is assuming that the original hormone has diffused away from its extracellular receptor.

Specificity of the cAMP system has two aspects. Specificity to the stimulatory hormone lies in the receptor for the hormone, on the extracellular surface of the membrane. Specificity with respect to the physiological response depends on the type of cell, and what target proteins are being phosphorylated and turned on or off inside the cell.

One way to remember what the effects will be when the cAMP system is stimulated is to consider cAMP as a "hunger signal." In this sense, cAMP activates those processes that produce energy, such as the *glycogen phosphorylase* cascade of liver, or lipolysis in adipose tissue. cAMP inhibits those processes that store energy, such as glycogen synthesis (via inactivation of *glycogen synthase*), or fatty acid synthesis (via inactivation of *acetyl CoA carboxylase*). All this is accomplished through phosphorylation of the appropriate enzymes.

2. *The Phosphoinositide Cascade* is used with some other peptide hormone systems, especially those that are calcium-dependent. Among these are α_1-adrenergics, angiotensin II, and acetylcholine (muscarinic). Other effects include glycogenolysis in liver and smooth muscle contraction.

Components of the system are (a) four proteins: the extracellular **receptor**, a **G-protein**, *phospholipase C*, and *protein kinase C*; (b) two second messengers: **inositol-1,4,5-trisphosphate** (IP_3) and **diacylglycerol** (**DAG**); and (c) **calcium ion**. In this context, Ca^{++} may be considered a *third* messenger: its level responds not only to IP_3 but also to cAMP.

Operation of the system depends on cleavage of the membrane lipid, phosphatidylinositol-4,5-bisphosphate (PIP_2) by *phospholipase C* to produce both second messengers in one step (Figure 9-6). Activation of the enzyme is linked to the extracellular hormone receptor by a G-protein, as in the cAMP system. But the production of dual second messengers leads to more complex consequences: (a) DAG activates *Protein Kinase C*, which goes on to phosphorylate target proteins; (b) IP_3 opens calcium channels, releasing Ca^{++} from storage in the ER. Ca^{++}, via calmodulin, a Ca^{++}-binding protein, then activates another set of target proteins. There is also some crossover in that Ca^{++} can further activate protein kinase C. These processes are illustrated in Figure 9-7.

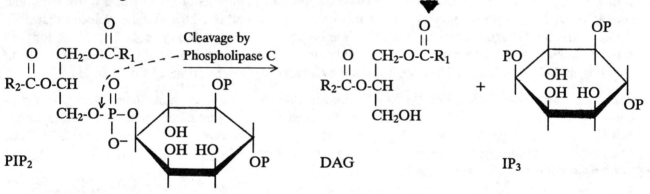

Figure 9-6. Conversion of Phosphatidylinositol-4,5-bisphosphate to DAG and IP_3.

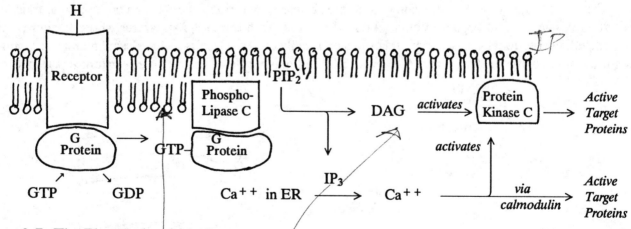

Figure 9-7. The Phosphoinositide System.

In muscle, IP_3 causes release of stored Ca^{++}, which, in turn, triggers contraction. Ca^{++} also further activates phosphorylase kinase, and stimulates the kinase that inactivates glycogen synthase. Thus contraction and glycogen metabolism are coordinated.

Turning the system off is accomplished by a phosphatase (*IP₃ase*) which converts IP$_3$ to IP$_2$. The latter is further metabolized to free inositol, which is then recycled. (Li$^+$ blocks removal of the last phosphate, interfering with re-use of inositol).

3. **Insulin** also binds to a transmembrane receptor, but its method of action is not well understood. It does not work through cAMP or phosphoinositides (though it may modulate the effects of these other systems), but the receptor itself is a protein kinase of a different sort. Activation by insulin results in stimulation of a tyrosine kinase activity on the cytoplasmic side of the receptor, which phosphorylates tyrosine residues in selected proteins. The target proteins include itself, which it autophosphorylates. How this translates into uptake of glucose and other nutrients is not clear. Internalization of the receptor occurs, which may be part of the action, or may simply be a way to control receptor concentration. Related hormone systems, eg., those involving the receptor for epidermal growth factor (EGF) and that for the insulin-like growth factor I (IGF-I) also have tyrosine kinase activity, as do certain oncogene products (eg., PDGF).

F. Other Membrane Phenomena

1. *Gap junctions* provide a means for direct communication between the cytoplasm of one cell and that of another. They consist of units called **connexons**, composed of subunits of **connexin**, a transmembrane protein. A channel through the middle is large enough to pass ions and small hydrophilic molecules (such as metabolites) directly through both membranes of adjoining cells, from cytosol to cytosol. These junctions occur in large numbers, and are important for cell-cell communication, nourishment of cells that are far from blood vessels, and differentiation. They can be closed by increased concentrations of Ca^{++} or H$^+$.

2. *Intercellular communication* and various *recognition phenomena* are often mediated through **gangliosides** and other carbohydrate-rich features of the extracellular leaflet of the membrane. Examples are **contact inhibition**, some **antigenic activities** (eg., ABO blood-group substances), and **receptor functions**. The latter can even be a feature of the action of certain toxins: **cholera toxin** binds specifically to the ganglioside G$_{M1}$, with eventual irreversible activation of adenylate cyclase.

3. *Synthesis of membranes* occurs only on pre-existing membranes. Since they are asymmetric, it is essential that new membranes are produced so that structural features such as carbohydrate residues are on the extracellular side, and directional functions such as ion pumps are properly oriented.

Synthesis of membrane proteins begins on membrane-bound ribosomes (rough endoplasmic reticulum), where glycosylation also begins. Vesicles bud off and fuse with the Golgi apparatus where further processing, including additional glycosylation, takes place. Finally the vesicles fuse with plasma membrane, preserving their original asymmetry (Figure 9-8), in which the luminal side of the ER is equivalent to the outside of the cell.

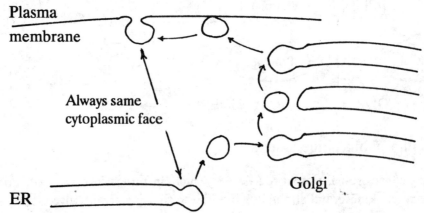

Figure 9-8. Synthesis of Membranes, Showing Preservation of Asymmetry.

IV. REVIEW QUESTIONS ON MEMBRANES

DIRECTIONS: For each of the following multiple-choice questions (1 - 10), choose the ONE BEST answer.

1. Biological membranes do all of the following EXCEPT:

A. freely pass small ions such as H^+.
B. freely pass gas molecules such as O_2.
C. freely pass small lipophilic molecules such as $CHCl_3$.
D. constitute a barrier to the passage of metabolites such as glucose-6-phosphate.
E. create ionic gradients.

2. All of the following contain glycerol EXCEPT:

A. lecithin.
B. sphingomyelin.
C. phosphatidyl ethanolamine.
D. phosphatidyl inositol.
E. cardiolipin.

3. The stability of membranes stems from:

A. ionic interactions between polar head-groups of amphipathic lipids and water.
B. ionic interactions between charged R-groups of proteins and water.
C. steric hindrance among hydrophobic portions of amphipathic lipids.
D. covalent bonds between lipid and protein.
E. hydrophobic repulsion of lipid chains by water, and other non-covalent interactions.

4. The side-chains of amino acids in the transmembrane domains of intrinsic proteins are likely to be rich in:

A. alanine and aspartate.
B. lysine and leucine.
C. glutamate and glutamine.
D. valine and isoleucine.
E. histidine and arginine.

5. Biological membranes contain all of the following EXCEPT:

A. cholesterol.
B. phosphatidyl choline.
C. triacylglycerols.
D. proteins.
E. sphingolipids.

6. A hormone whose action involves tyrosine kinase activity is:

A. epinephrine.
B. glucagon.
C. ACTH.
D. acetylcholine.
E. insulin.

7. Inactivation of the phosphoinositide system may occur as a result of all of the following EXCEPT:

A. diffusion of the stimulating hormone away from its receptor.
B. phosphorylation of the G protein.
C. GTPase activity of the G protein.
D. conversion of IP_3 to IP_2.
E. dephosphorylation of phosphorylated proteins by a phosphatase.

8. The presence of cholesterol in a membrane

A. increases fluidity at both low and high temperatures.
B. decreases fluidity at both low and high temperatures.
C. increases fluidity at low temperatures and decreases it at high temperatures.
D. decreases fluidity at low temperatures and increases it at high temperatures.
E. has no effect on fluidity.

9. Fluidity of membranes

A. is favored by longer fatty acid chains in the lipids.
B. is favored by a greater degree of saturation in the lipids.
C. is favored by the presence of double bonds in the lipids.
D. is insufficient to allow lateral movement of proteins within the plane of the leaflet.
E. allows crossing over of phospholipid molecules from one leaflet to the other.

10. Hormones that act primarily through the cAMP system include all of the following EXCEPT:

A. epinephrine
B. α_1-adrenergics
C. β-adrenergics
D. glucagon
E. ACTH

MATCHING: For each set of questions, choose the ONE BEST answer from the list of lettered options above it. An answer may be used one or more times, or not at all.

Questions 11 - 21:

A. Extracellular receptor.
B. G protein.
C. Phospholipase C.
D. Phosphatidylinositol-4,5-bisphosphate (PIP_2).
E. Inositol-1,4,5-trisphosphate (IP_3).

F. Diacylglycerol (DAG).
G. Protein kinase C.
H. Calcium ion.
I. Calmodulin.
J. Phosphodiesterase.

11. In a sense, a third messenger.

12. A transmembrane protein.

13. Second messenger that opens calcium channels.

14. A calcium-binding protein.

15. A GTPase.

16. Phosphorylates target proteins.

17. Second messenger that activates protein kinase C.

18. Enzyme that produces two second messengers in one step.

19. Substrate whose cleavage produces two second messengers in one step.

20. Binds acetylcholine.

21. NOT a part of the phosphoinositide cascade.

Questions 22 - 30:

   A. Simple diffusion.
   B. A facilitated transport system that is a uniport.
   C. A facilitated transport system that is an antiport.
   D. An active transport system that is a symport.

   E. Receptor-mediated endocytosis.
   F. An F-type ATPase.
   G. A P-type ATPase that is a uniport.
   H. A P-type ATPase that is an antiport.

22. The Cl^- - HCO_3^- transporter of red cell membranes.

23. Movement of O_2 across the inner mitochondrial membrane.

24. The ATP-generating system of mitochondrial oxidative phosphorylation.

25. Absorption of glucose from intestinal lumen into brush border cells.

26. Uptake of LDL.

27. Glucose permease of red cell membranes.

28. The Na^+ - K^+ pump.

29. Calcium ion pump of muscle.

30. Valinomycin.

Questions 31 - 36:

   A. Interaction between a hormone and its extracellular receptor.
   B. Transmission of a signal across the membrane.
   C. Passage of the hormone across the membrane.
   D. G Protein.

   E. Adenylate cyclase.
   F. cAMP
   G. Phosphorylated proteins.
   H. Phosphodiesterase.

31. Intermediary between transmembrane protein and the enzyme that produces a second messenger.

32. Specificity as to whether a cell will respond to a particular hormonal stimulus.

33. One way to turn the system off.

34. NOT a part of the system by which glucagon regulates glycogen metabolism in liver.

35. Specificity as to type of physiological response.

36. Second messenger.

V. ANSWERS TO QUESTIONS ON MEMBRANES

1. A	10. B	19. D	28. H
2. B	11. H	20. A	29. G
3. E	12. A	21. J	30. B
4. D	13. E	22. C	31. D
5. C	14. I	23. A	32. A
6. E	15. B	24. F	33. H
7. B	16. G	25. D	34. C
8. C	17. F	26. E	35. G
9. C	18. C	27. B	36. F

10. NUTRITION

Thomas Briggs

I. MAJOR NUTRIENTS

A. Energy Nutrition

The unit generally used is the **Kilocalorie** (Kcal, popularly but incorrectly also known as the Large Calorie, Cal): amount of energy needed to raise the temperature of 1 Kg water from 14.5° to 15.5°C. The **Megajoule** is gaining some acceptance: 1 MJ = 239 Kcal; 1 Kcal = 4.2 KJ. Ten MJ is roughly a day's supply of energy, at a moderate level of activity.

Caloric yields from the metabolism of fuels are:

Carbohydrate	4 Kcal/g	Fat	9 Kcal/g
Protein	4 Kcal/g	Ethanol	7 Kcal/g

Respiratory Quotient (RQ) is the volume of CO_2 produced / volume of O_2 consumed. This varies depending on the type of fuel being oxidized: 1.0 for carbohydrate, to 0.7 for fat.

Basal Metabolic Rate (BMR) is the rate of oxygen consumption, or equivalent heat production, of an awake individual, at rest, who has not eaten for at least 12 hours. To compensate for size differences, the BMR is usually expressed per unit surface area. A typical figure might be 35 Kcal/hr/m². It is higher in children, males, hyperthyroidism, etc. Resting energy expenditure (REE) is less precisely defined, but differs little from BMR.

Thermic Effect of Food (formerly known as specific dynamic action, SDA) is an increase in metabolic rate after eating, resulting from the energy cost of metabolism, particularly of protein, and may be about 6% on a normal mixed diet.

The daily requirement of energy varies greatly, depending on BMR and especially on muscular activity. For a sedentary 70 Kg man a typical value may be 2400 Kcal, but very strenuous activity could double this.

B. Fuels

Carbohydrates: the major dietary form is the polysaccharide, **starch**. Important oligosaccharides include **sucrose, lactose**. Monosaccharides occur increasingly in the US, due to the widespread use of high-**fructose** corn syrup (which may also contain glucose). Carbohydrates function mainly as fuels, with minor amounts incorporated into glycolipids and glycoproteins.

Fats (lipids): **triacylglycerols** (TG) are the major dietary form. The dietarily essential lipids are the unsaturated fatty acids **linoleic** (18:2[9,12]) and **linolenic** (18:3[9,12,15]) **acids**. Recommended dietary allowances for these have not been established.

Although lipids have a major function as fuels and in the storage of energy, they also serve in the structure of membranes (phospholipids, cholesterol), in the formation of prostaglandins (poly-unsaturated fatty acids), as other hormones (steroids), as thermal insulators, and even for decoration (deposits of fatty tissue).

The amount and type of fat in the diet influence the levels of cholesterol in the body. A diet with a high ratio of polyunsaturated to saturated fatty acids, but low in total fat, tends to promote low levels of cholesterol in the blood.

C. Protein

Though proteins consumed in excess of the daily requirement are simply used as fuels, the most significant dietary function is as a source of amino nitrogen for synthesis of body constituents.

Ten amino acids are considered dietarily essential because the human cannot synthesize the carbon skeleton (mnemonic: PVT TIM HALL):

Phenylalanine	Threonine	Histidine
Valine	Isoleucine	Arginine*
Tryptophan	Methionine	Leucine
		Lysine

*A dietary requirement has not been rigorously established for arginine in the <u>adult</u> human.

The quality of dietary proteins depends on (1) digestibility and (2) the content of essential amino acids. The biological value (BV) depends mainly on amino acid composition. A more useful index, **Net Protein Utilization** (NPU), considers both (1) and (2) above. The NPU of human milk is 95%; of cow's milk, 81%; of wheat protein, 49%; of corn protein, 36%. Cooking can increase the NPU, as can mixing proteins with complementary amino acid compositions. This is especially important in vegetarian diets.

The **nitrogen balance** is zero when intake just equals loss (as urea, digestive losses, etc.). For this to occur, all essential amino acids must be present in the diet in sufficient quantities. If even one is deficient, negative balance results. Growth, convalescence, and pregnancy are accompanied by positive nitrogen balance; illness, fever, starvation by negative balance.

The Recommended Dietary Allowance (RDA, see below) for protein consumed by adult males and females on a typical US diet is **0.8 g/Kg/day**.

D. Other

Fiber: non-digested (especially plant) material such as cellulose, lignin, etc. Though not generally regarded as a dietary essential, fiber nevertheless has beneficial effects on digestion. Through its *bulking* action, it promotes good mechanical functioning of the digestive tract; by a *speeding* action, it reduces the transit time for intestinal contents and therefore the time available for bacteria to produce possible carcinogens; by *binding* bile salts it is thought to increase the turnover of the bile salt pool and thus to promote excretion of cholesterol through increased conversion of cholesterol to bile acids. Fiber may, however, decrease absorption of certain nutrients, such as iron and calcium.

Water: requirement is highly variable. It functions as a *solvent* for components of blood and tissues, and as a medium for excretion of wastes. Another important use is in *regulation of body temperature*.

E. Health-Related Issues

Obesity is a pervasive health problem related to energy intake because it predisposes to numerous conditions including cardiovascular disease, diabetes, and gallbladder disease.

On a world-wide basis, **malnutrition** is also a severe problem. There are two "pure" manifestations of malnutrition: *marasmus*, where total caloric intake is deficient, and *kwashiorkor*, where calories are sufficient but intake of protein is deficient. A real situation may be a combination of these.

For diseases related to deficiency of specific nutrients, see the appropriate sections to follow.

F. Dietary Recommendations

Recommended Dietary Allowance (RDA): issued for each nutrient by the Food and Nutrition Board of the National Academy of Sciences, and periodically revised in the light of new knowledge. Defined as the amount of a nutrient which, if consumed by every member of a population, will keep nearly everyone in good health, it can also have political overtones as some public health measures may be tied to it. Since people vary greatly in their dietary requirements, no one figure is applicable to all. The RDA is based on an average requirement (which may not be known with precision), two standard deviations are added, and often a safety factor besides, so that it is a generous excess for most people. Detailed tables are subdivided according to gender, age, etc. It should not be confused with an individual's minimum requirement.

RDA values for individual nutrients are given in the following sections that discuss each.

Caloric Balance: The typical diet in the US derives about 35 - 40% of its caloric value from fat. This is considered high; it is now recommended that no more than 30% of calories come from fat, and of these, a maximum of one-third from saturated fat. If about 10% of calories are derived from protein, then the remainder, at least 60%, should be provided by carbohydrate.

II. MICRONUTRIENTS

A. Vitamins

1. *Fat-Soluble*: isoprenoid derivatives with varying degrees of unsaturation.

    a. **Vitamin A**: derived from a pro-vitamin, β-carotene, a yellow plant pigment which is cleaved to vitamin A by an enzyme in intestinal mucosa.

    i. *Structure*: occurs in three forms: **retinol**, vitamin A aldehyde (**retinal**), vitamin A acid (**retinoic acid**). The structure shown is that of retinol.

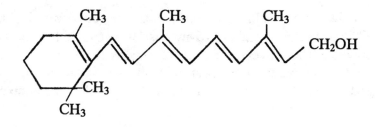

    ii. *Function*: not known with precision except in vision, in which retinal is combined with the protein, opsin, to form the visual pigment, **rhodopsin**. The process of vision depends on the reversible conversion of **11-*cis*-retinal** to **all-*trans*-retinal**. Further functions of vitamin A (for which a dietary supply of any form will usually suffice) are in differentiation of epithelial cells, growth, reproduction, and in the immune system. It seems to be particularly important in fetal development and may function, along with a receptor, as a transcription factor or in other ways to influence genetic expression.

    iii. *Occurrence*: β-**carotene** in green and yellow vegetables, carrots, pumpkin, cantaloupe melons. Vitamin A itself occurs only in animal products, especially butter (but often there by fortification), eggs, fish liver oil, liver (polar bear liver may contain toxic amounts).

iv. *Deficiency*: night blindness is an early warning. This may advance to keratinization of epithelial tissues, especially of the eye, causing **xerophthalmia**, a form of blindness which is of major concern in some areas.

v. *Requirement*:[*] **0.8-1.0 mg** (as retinol), or six times that amount as β-carotene, since the conversion and/or absorption is variable and incomplete. Sustained high intake of vitamin A causes many toxic symptoms and abnormalities including birth defects. Carotenoids are not known to be toxic.

b. **Vitamin D**

i. *Structure*: cholecalciferol is derived from **7-dehydrocholesterol** through cleavage by UV light of the 9,10-bond, then rotation of ring A 180° around the 6,7-bond (Figure 10-1).

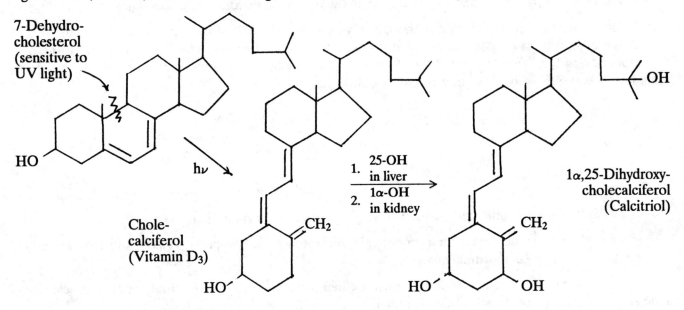

Figure 10-1. Formation and Metabolism of Vitamin D

ii. *Function*: Cholecalciferol is like a prohormone. The active form is generated by (1) insertion of a hydroxyl group at C-25 by the liver, and (2) 1α-hydroxylation by a kidney enzyme whose activity is enhanced by parathyroid hormone. The active **1α,25-dihydroxycholecalciferol (calcitriol)** acts in a manner similar to that of the steroid hormones, to cause synthesis of a calcium-binding protein by intestinal cells. It may also be active in bone and kidney. The effect is to raise the level of blood **calcium** and to promote mineralization of bone.

iii. *Occurrence* in foods is not necessary if a person has a sufficient exposure to sunlight. Otherwise, dairy products, especially fortified milk, and fish liver oils are good sources.

iv. *Deficiency*: **rickets**. Adult form is **osteomalacia**. Primary deficiency is rare in developed countries, but is sometimes found in individuals with kidney failure.

v. *Requirement*: **5-10 μg** (in the absence of sunlight). Excess can cause irreversible damage; the toxic level may be only 5 times the RDA.

---

[*] In this chapter, requirements are the RDA for adults, as presented in the Tenth Edition, *Recommended Dietary Allowances*, National Academy Press, Washington, 1989.

c. **Vitamin E: tocopherols,** α, β, etc., depending on the length of the isoprenoid side-chain.

    i. *Structure*:

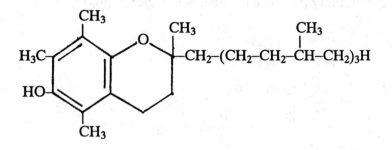

    ii. *Functions* as an **antioxidant**, probably by virtue of its ability to act as a trap for free radicals.

    iii. *Occurs* widely, especially in vegetable oils.

    iv. *Deficiency*: in animals, sterility and muscular dystrophy. In humans, anemia and neurological disorders may occur rarely, in association with malabsorption syndromes.

    v. *Requirement*: **8-10** mg/day, but may be greater with increased consumption of polyunsaturated oils. These, however, are also good sources of vitamin E.

d. **Vitamin K (phylloquinone)**

    i. *Structure*: a family of substituted naphthoquinones. Unsubstituted menadione is also active.

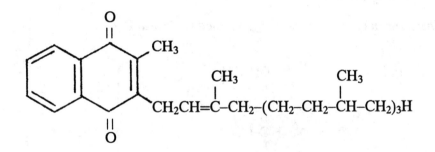

    ii. *Function*: needed for the carboxylation of glutamyl residues to produce γ-**carboxyglutamic acid** in active blood **clotting factors prothrombin (II), VII, IX, X** and **proteins C** and **S**. These γ-carboxyglutamic acids are able to chelate calcium. They also occur in other tissues, but with unknown function.

    iii. *Occurrence*: ubiquitous in foods, especially green leafy vegetables, and also synthesized by intestinal bacteria.

    iv. *Deficiency* is rare but can occur in the newborn infant and in malabsorption syndromes, causing **hemorrhage** due to hypoprothrombinemia. **Dicoumarol** and related compounds are antagonists of Vitamin K and are used to prevent thrombosis.

    v. *Requirement*: **1 μg/Kg** body weight, or **60-80 μg/day**. Some bacterially-synthesized K can be absorbed, but not in sufficient amounts to replace the need for dietary K.

2. *Water-Soluble, Energy Releasing*

  a. **Thiamin (Vitamin B$_1$)**

    i. *Structure*: a substituted pyrimidine joined to a substituted thiazole.

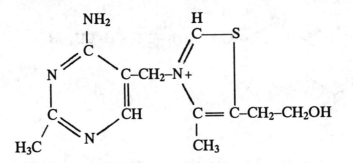

    ii. *Cofactor form and function*: **Thiamin pyrophosphate (TPP).** Part of the *pyruvate dehydrogenase* complex, TPP functions in the **decarboxylation** of pyruvate to form acetyl CoA. It is also part of *α-ketoglutarate dehydrogenase* and of *transketolase*.

    iii. *Occurrence*: meat, liver, whole grains, legumes.

    iv. *Deficiency*: the classical disease is **beriberi**. Alcoholics may have a multiple B-vitamin deficiency known as **Wernicke's disease.**

    v. *Requirement*: **1-1.5 mg/day**, sometimes expressed as as a function of daily energy expenditure — 0.5 mg / 1000 Kcal.

  b. **Riboflavin (Vitamin B$_2$)**

    i. *Structure*: a heterotricyclic system joined to ribitol.

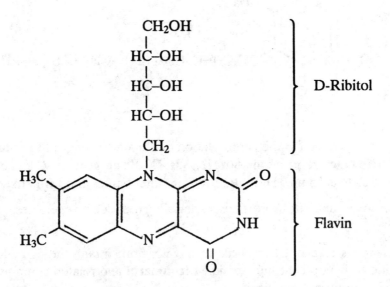

    ii. *Cofactor form and function*: **flavin mononucleotide** or **FMN** (riboflavin phosphate); **flavin adenine dinucleotide** or **FAD** (a combined nucleotide with AMP). These, as part of flavoproteins, act in **transfer of hydrogen and electrons** from NAD to CoQ, and from succinate, from acyl CoA in β-oxidation, etc. FAD is part of *pyruvate* and *α-ketoglutarate dehydrogenases*. Derivatives of riboflavin are also involved in functioning of B$_6$ and niacin.

iii. *Occurrence*: milk, liver, meat, green vegetables. Cooking or exposure to light tends to destroy riboflavin.

iv. *Deficiency*: lesions of the lips, skin, genitalia.

v. *Requirement*: **1-1.5 mg/day** or 0.6 mg/1000 Kcal.

c. **Nicotinamide, Nicotinic Acid (Niacin)**

i. *Structure*: a substituted pyridine.

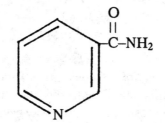

ii. *Cofactor form and function*: **Nicotinamide adenine dinucleotide** or **NAD** (nicotinamide + adenine + 2 ribose + 2 phosphate), **NADP** (NAD + a third phosphate). These act as **carriers of hydrogen and electrons** in a multitude of dehydrogenase reactions. In general, NAD acts in catabolism while NADP acts in synthetic reactions. NADPH and O_2 are used by mixed function oxidases, particularly in metabolism of drugs and in various hydroxylations.

iii. *Occurrence*: meat, liver, peanuts, legumes, whole grains. Nicotinate can be biosynthesized from tryptophan, but inefficiently (1 mg from 60 mg). Corn is a poor source of both niacin and tryptophan.

iv. The *deficiency* disease is **pellagra**, characterized by diarrhea, dermatitis, dementia, and death ("4 D's"). Pellagra is of historical interest as it used to occur in the US over a geographic area where the population was particularly dependent on corn as a dietary staple.

v. *Requirement*: **15-20 mg/day**, or 6.6 mg/1000 Kcal.

d. **Pyridoxine (Vitamin B$_6$)**

i. *Structure*: a substituted pyridine. Also occurs as the aldehyde and amine forms.

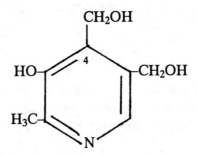

ii. *Cofactor form and function*: **pyridoxal phosphate** and **pyridoxamine phosphate**. These act as coenzymes of *amino acid transaminases*, some *decarboxylases*, *glycogen phosphorylase*, *ALA synthase*, etc. They reversibly form a Schiff base with amino groups.

iii. *Occurrence*: fish, chicken, liver, whole grains.

iv. *Deficiency*: convulsions, oxalate kidney stones. Deficiency can be caused by treatment with the drug isoniazid, which forms a rapidly-excreted hydrazone with pyridoxal.

v. *Requirement*: **2 mg/day**. Prolonged ingestion of megadoses may cause neurological effects.

e. Pantothenic Acid

i. *Structure*: pantoic acid + β-alanine.

$$\text{HO--CH}_2\text{--}\underset{\underset{\text{CH}_3}{|}}{\overset{\overset{\text{H}_3\text{C}}{|}}{\text{C}}}\text{--}\overset{\overset{\text{OH}}{|}}{\text{CH}}\text{--}\underset{\underset{\text{O}}{\|}}{\text{C}}\text{--}\overset{\overset{\text{H}}{|}}{\text{N}}\text{--CH}_2\text{--CH}_2\text{--COOH}$$

ii. *Cofactor form and function*: **Coenzyme A** (phosphoadenosine diphosphate attached to the pantoic acid moiety, + thioethanolamine attached to the β-alanine moiety). The terminal -SH combines with acyl groups to form a "high-energy" thioester; this is the way acyl compounds are activated before undergoing metabolism; example: acetyl CoA. The pantothenic acid and thioethanolamine are also part of *acyl carrier protein*, in which the -SH carries the growing chain during fatty acid synthesis.

iii. *Occurrence*: ubiquitous in foods, especially animal tissues, whole grains, legumes. Some may be produced by intestinal microflora.

iv. *Deficiency*: extremely rare.

v. *Requirement*: **4-7 mg/day** ("safe and adequate," no RDA has been set).

f. Biotin

i. *Structure*:

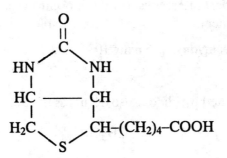

ii. *Cofactor form and function*: biotin acts in several **carboxylations**, being a part of *pyruvate carboxylase* and *acetyl CoA carboxylase*, among others. It is linked to its enzymes by an ε-amino group of lysine, forming a swinging arm which enables it to transfer a -COOH from one active site to another in an enzyme complex.

iii. *Occurrence*: in many foods, and is made by intestinal bacteria in amounts possibly sufficient to satisfy needs.

iv. *Deficiency*: very rare; can be induced by consumption of large amounts of raw egg whites which contain a glycoprotein, **avidin**, which binds biotin and renders it unabsorbable.

v. *Requirement*: **30-100 μg / day** (estimated safe and adequate). No RDA has been set, as knowledge is incomplete regarding requirement *vs.* relative amounts provided by diet and intestinal flora.

3. Water-Soluble, Hematopoietic

a. Folic Acid (pteroyl glutamic acid, folacin)

i. structure

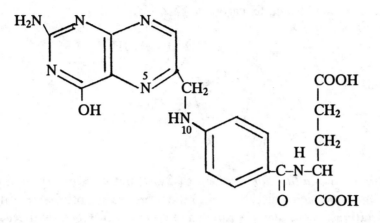

ii. *Cofactor form and function*: **tetrahydrofolic acid (THFA)**, with several glutamic acids attached. Active in metabolism of **one-carbon** units, THFA carries a C-1 fragment, at various states of oxidation, between N^5 and N^{10} (Chapter 5). THFA is also essential in **purine** biosynthesis (Chapter 11), in the reversible transformation of **serine** to glycine, and in other aspects of one-carbon metabolism. The transported form of the coenzyme is N^5-methyl THFA; regeneration to active THFA requires adenosyl cobalamin, one of the forms of Vitamin B_{12}.

Certain analogues of folic acid are useful as anticancer agents: **aminopterin**, **amethopterin**, and **methotrexate** (Chapter 11).

iii. *Occurrence*: kidney, liver, dark green leafy vegetables (hence the name).

iv. *Deficiency*: megaloblastic bone marrow and macrocytic anemia.

v. *Requirement*: **200 μg/day**. Folate and the drug phenytoin compete for absorption. There is some potential for toxicity from excessive intake.

b. Cobalamin (Vitamin B_{12})

i. *Structure*: a **corrin** derivative, a complex tetrapyrrole related to the porphyrins but lacking one of the methenyl bridges, and containing **cobalt**.

ii. *Cofactor form and function*: as the **deoxyadenosine** or **methyl** derivative. (Cyanocobalamin is a form often obtained from isolation procedures). B_{12} functions in two areas: (1) regeneration of **active tetrahydrofolic acid** (during which a methyl group is transferred to homocysteine to form methionine); (2) *mutase* reactions, as in the conversion of methylmalonyl CoA to succinyl CoA.

iii. *Occurrence*: made by certain microorganisms, and obtained in the diet only from foods of animal origin, especially meat and dairy products. Absent in plant foods.

iv. *Deficiency*: all the symptoms of folate deficiency plus neurological abnormalities. In **pernicious anemia**, the deficiency is really one of **intrinsic factor**, a glycoprotein produced by the stomach, which is necessary for ileal absorption of B_{12}. **Methylmalonic aciduria** is one effect. Dietary deficiency may rarely occur in strict vegetarians.

v. *Requirement*: **2 µg/day**. But unlike other B-vitamins, B$_{12}$ can be stored; the liver can hold a several years' supply. It also undergoes an efficient enterohepatic circulation, which greatly prolongs its lifetime in the body.

4. *Other Water-Soluble*: *Ascorbic Acid (Vitamin C)*

i. *Structure*: related to the six-carbon sugars.

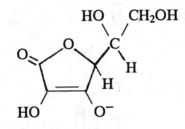

ii. *Cofactor form and function*: cofactor (if any) not known. Vitamin C functions as a **reducing agent**, and is important in reducing dietary iron to the more easily absorbable ferrous form. It also is essential in some **hydroxylations**, especially of **proline** and **lysine** in **collagen** synthesis. It also probably has a function (unknown) in the metabolism of the adrenal cortex.

iii. *Occurrence*: citrus fruits, tomatoes, other fruits and vegetables.

iv. *Deficiency*: The classical disease is **scurvy**, which is particularly a disease of collagen formation: poor wound healing, fragile bones, loosening of teeth, hemorrhage. Though blood levels of ascorbic acid decline rapidly on a deficient diet, scurvy usually takes many weeks to develop. Some authorities maintain that a state of deficiency short of outright scurvy is a very common condition.

v. *Requirement*: **60 mg/day**. Since the turnover of vitamin C is increased in cigarette smokers, it is recommended that these should ingest 100 mg/day.

B. Minerals

1. *Electrolytes*

a. **Sodium**: The major **extracellular cation**, sodium functions in osmotic, water, and acid-base balance. It is ubiquitous in foods, especially those of animal origin and in preserved, prepared, and processed foods. A deficiency is uncommon, but can occur in an unacclimatized person through prolonged heavy sweating. An excess is widespread and insidious, and may cause **hypertension** in susceptible individuals. Estimated requirement: **0.5 g/day**, but not over 2.4 g/day (equivalent to 1.3 g and 6 g NaCl, respectively).

b. **Potassium**: The major **intracellular cation**, potassium also functions in osmotic, water, and acid-base balance. It occurs widely in (unprocessed) foods, especially those of plant origin. Deficiency may occur after prolonged vomiting, in diarrhea, diabetics, and in those on diuretics. Cardiac abnormalities, possibly fatal, may ensue. Excretion of sodium and potassium by the kidney is regulated by the **renin-angiotensin system**. Estimated requirement: **1.6-2.0 g/day**.

c. **Chloride**: The principal **anion** for Na$^+$ and K$^+$, it occurs everywhere and deficiency is not a problem. Estimated requirement: **1.8 g/day**.

2. *Other Major Minerals*

a. **Calcium**: the major inorganic element, calcium is 1.5-2% of the body's mass. Most is in the **skeleton**, but calcium has vital physiological roles in **muscle contraction, blood clotting,** as an intracellular **messenger**, and in many other ways. The level of blood calcium is regulated by **parathyroid hormone** (raises level), **vitamin D** (promotes absorption in intestine), **calcitonin** (lowers level), and by other hormones. Bone serves as a buffer or reservoir.

Bone density increases during the first $2\,^1/_2$ decades of life, during which time it is most important that adequate calcium intake be maintained. Later, losses of calcium predominate which may lead to osteomalacia, especially among those whose maximal bone density is low due to earlier inadequate intake. Post-menopausal women are also at particular risk.

Dairy products are good sources, as are cruciferous vegetables and some calcium-prepared foods. A primary dietary deficiency in adults is rare, but can occur secondary to vitamin D deficiency, and in women after multiple pregnancies and lactation. It sometimes is seen in a newborn on cows' milk (which has plenty of calcium but a low Ca/P ratio, which may make regulation difficult).

The adult requirement is **0.8-1.2 g/day** but adaptation or high bioavailability may make the true need much less. Post-menopausal women and the aged may have an increased requirement. It is not clear to what extent, if at all, calcium supplementation reverses loss of bone in the older population.

b. **Phosphorus**: an important **intracellular anion** (as phosphate), phosphorus is about 1% of the body's mass, occurring mostly as a constituent of **bone**. Chemically it has major functions as a component of **nucleic acids**, many **nucleotides**, and **phosphate esters** of sugars and other intermediates. A dietary deficiency is virtually unknown, but low levels may occur in the diabetic and in individuals receiving prolonged antacid treatment with aluminum hydroxide. Requirement: **0.8-1.2 g/day**.

c. **Magnesium**: after potassium, an important intracellular cation. It occurs in **bone**, and also functions in many enzyme activities, especially those involving **ATP**. Most foods of vegetable origin are good sources, especially whole grains. Deficiency may occur in alcoholics. Requirement: **280-350 mg/day**.

3. *Iron*: Functioning in oxygen transport and storage, iron is part of the heme proteins *hemoglobin* and *myoglobin*. It is also part of the *cytochromes* of the electron transport chain. Other heme-containing enzymes are *catalase* and *peroxidase* which are involved in the metabolism of H_2O_2. The average adult has about 4 g, $^2/_3$ of which is in hemoglobin, and $^1/_4$ in storage, particularly as liver ferritin. The phases of iron metabolism are outlined below:

a. *Absorption* of dietary iron: highly variable. Only about 10% or less (1-2 mg/day) of non-heme iron may be absorbed, as the **reduced**, or **ferrous** form; heme iron is more efficiently absorbed, up to 40%.

b. *Regulation* is at the level of **absorption**. According to one proposal, in response to adequate body iron stores, **intestinal cells** produce a **ferritin trap** which prevents further passage of absorbed iron into the bloodstream before the cells slough off. When stores are low and there is a need for iron, ferritin is not produced by intestinal cells, and iron that is taken in is allowed to pass through into the circulation.

c. *Oxidation*: ferrous iron is oxidized to the ferric form by *ferroxidase (ceruloplasmin)*, a **copper**-containing enzyme.

d. *Transport*: iron is tightly bound by *transferrin*, two ferric atoms per molecule.

e. *Storage*: mostly in *ferritin* and *hemosiderin* of liver, some in bone marrow. Many ferric ions per molecule.

f. *Excretion*: virtually non-existent. The only way for the body to be significantly depleted of iron is by **bleeding** or **childbearing**. Iron overload may occur in some alcoholics, recipients of blood transfusions for hemolytic anemia, and overzealous practitioners of self-medication. Hemochromatosis is an inborn error in which too much dietary iron is absorbed, leading to excessive accumulation.

g. *Occurrence*: meat (heme iron is well-absorbed), egg yolk, legumes. Milk and spinach are poor sources. Simultaneous intake of **vitamin C** enhances the efficiency of absorption by helping to keep the iron in the reduced state. Substances that inhibit absorption include certain fibers, antacids, and polyphenols of tea.

h. *Deficiency*: iron-deficiency anemia is relatively common since dietary iron is inefficiently absorbed. It is most often seen in young children, early adolescence, young mothers, and in persons with chronic loss of blood.

i. *Requirement*: **10 mg/day** (**12-15** for teenagers; **15** for women of childbearing age).

4. *Trace minerals* with established RDA

a. **Iodide**: required for **thyroid** function — T_4 and T_3. Found in sea foods, iodized salt. Requirement: **150 μg/day**. Deficiency is manifested as endemic **goiter**.

b. **Selenium**: functions in *glutathione peroxidase*, which, as an **antioxidant**, is complementary to vitamin E. Requirement: **55-70 μg/day**. Toxic in the mg/day range.

c. **Zinc**: important in the immune system and as a cofactor for many enzymes, eg., *carbonic anhydrase, alcohol dehydrogenase, nucleic acid polymerases*. Deficiency is rare, but has been seen in alcoholics and in cases of renal disease. Absorption is inhibited by some other minerals and bran. Requirement: **12-15 mg/day**.

5. *Other trace minerals* with estimated safe and adequate dietary intakes

a. **Chromium**: involved with the action of **insulin**. Estimated need: **50-200 μg/day**.

b. **Copper**: part of a number of enzymes, notably *cytochrome oxidase* and other oxidases including *ceruloplasmin*. Wilson's disease is an inborn error involving excessive deposition of copper in the liver. Estimated need: **1.5-3 mg/day**.

c. **Fluoride**: Probably not an essential element, but 1 ppm in the water supply is beneficial in helping to prevent **dental caries**. Estimated need: **1.5-4 mg/day**.

d. **Manganese**: widely distributed in vegetable foods, manganese participates in a variety of enzymic activities, such as *pyruvate carboxylase*. Estimated need: **2-5 mg/day**.

e. **Molybdenum**: required for *xanthine oxidase, aldehyde oxidase, sulfite oxidase*. Antagonistic to copper. Estimated need, easily furnished by diet: **75-250 μg/day**

6. *Miscellaneous*

a. **Carnitine**: can be synthesized by the human, and it occurs widely in foods of animal origin. Deficiency due to a heritable defect in its synthesis has been described.

b. **Choline**: widely present in foods, choline is a constituent of membranes as phosphatidyl choline or lecithin, and is required in the diet by some animals, as a source of methyl groups that can partially spare the methionine requirement. No dietary requirement has been shown in adult humans, though it may be essential in the diet of neonates.

c. **Cobalt**: the only need is as a component of **vitamin B_{12}**.

d. **Inositol**: important in membranes and in a second messenger system, inositol is provided by diet and synthesis. Deficiency has been produced in some animals, but has never been observed in humans.

III. SUMMARY OF NUTRIENT - FUNCTION ASSOCIATIONS

Vitamin	Cofactor Form	Association or Function
A	11-cis Retinal	Rhodopsin in vision
	Other forms	Growth, differentiation, epithelial tissues
D	$1\alpha,25$-Dihydroxycholecalciferol	Intestinal calcium-binding protein
E	none	Antioxidant
K	none	Carboxylation of clotting factors
Ascorbic acid (C)	none	Reducing agent, hydroxylations
Biotin	none	Carboxylations
Cobalamin (B_{12})	Deoxyadenosyl or methyl derivative	Mutases, recovery of active THFA
Folic acid	Tetrahydro derivative	Metabolism of 1-Carbon units
Niacin (B_3)	NAD	Dehydrogenases, catabolic
	NADP	Dehydrogenases, anabolic
Pantothenic acid	Coenzyme A	Activation of acyl groups
	Acyl Carrier protein	Growing chain in fatty acid synthesis
Pyridoxine (B_6)	Pyridoxal-P	Transaminases, decarboxylases
Riboflavin (B_2)	FMN, FAD	Dehydrogenases, catabolic
Thiamin (B_1)	Thiamin PP	Pyruvate, α-KG decarboxylases

Mineral	Association or Function
Calcium	Bone, coagulation, muscle contraction, second messenger
Chlorine	As chloride, anion for Na^+, K^+, etc.
Chromium	Potentiation of insulin
Cobalt	Component of vitamin B_{12}
Copper	Cytochrome oxidase, ferroxidase
Fluorine	Prevent dental caries
Iodine	Component of thyroid hormones
Iron	Hemoglobin, myoglobin, in cytochromes of e^- transport, catalase, peroxidases
Magnesium	In bone, kinase reactions
Manganese	Arginase, acetyl CoA carboxylase
Molybdenum	Xanthine oxidase, etc.
Phosphorus	As phosphate, in bone, DNA, RNA, nucleotides, organic phosphates
Potassium	Intracellular cation, osmotic, acid-base balance
Selenium	Glutathione peroxidase
Sodium	Extracellular cation, osmotic, acid-base balance
Zinc	Carbonic anhydrase, alcohol dehydrogenase, etc.

IV. REVIEW QUESTIONS ON NUTRITION

DIRECTIONS: For each of the following multiple-choice questions (1 - 26), choose the ONE BEST answer.

---

1. Which of the following can completely replace dietary methionine?

A. betaine
B. cysteine
C. homocysteine
D. vitamin B_6
E. ornithine.

2. Which of the following is most deficient in corn meal?

A. valine
B. methionine
C. threonine
D. tryptophan
E. leucine.

3. Which of the following is NOT a dietarily essential amino acid?

A. methionine
B. phenylalanine
C. leucine
D. isoleucine
E. serine.

4. In general, animal proteins are nutritionally better than vegetable proteins because:

A. though they have no important difference in amino acid composition, they are more easily digested than vegetable proteins
B. the content of essential amino acids more closely matches the body's needs
C. they contain sufficient carbohydrate residues
D. vegetable proteins lack asparagine
E. vegetable proteins contain factors which inhibit protein synthesis.

5. A zero nitrogen balance most likely occurs in:

A. a normally growing child
B. a normal adult
C. a child on a diet low in tyrosine
D. an adult convalescing after surgery
E. a fasting adult

6. Ferritin:

A. is a plasma protein which binds iron
B. is a muscle protein which oxidizes iron to the ferric state
C. is a protein which stores ferric ions
D. is involved in absorption of vitamin B_{12} from intestine
E. helps regulate excretion of iron in kidney.

7. Periodic injections of vitamin B_{12} can alleviate and prevent recurrence of the symptoms of pernicious anemia. These must be repeated monthly for:

A. 3 months
B. 1 year
C. 2 years
D. 5 years
E. a lifetime.

8. Absorption of iron from the intestine is increased when:

A. non-heme dietary iron is in the ferrous (Fe^{++}) rather than the ferric (Fe^{+++}) state
B. the diet contains a greater proportion of heme, rather than non-heme iron
C. ascorbic acid is taken in the same meal with non-heme iron
D. hemorrhage has occurred
E. all of the above.

9. Which of the following is impaired in biotin deficiency?

A. synthesis of ketone bodies from pyruvate
B. formation of lactate from glucose
C. synthesis of fatty acids
D. oxidation of fatty acids
E. elongation of fatty acids.

10. Long-standing biliary obstruction may cause malabsorption which can cause vitamin deficiency leading to:

A. osteomalacia
B. xerophthalmia
C. night blindness
D. a tendency to bleed
E. all the above.

11. Adenosyl cobalamin is involved in the conversion of:

A. malonyl CoA to acetyl CoA
B. α-ketoglutarate to succinyl CoA
C. Acetyl CoA to HMG CoA
D. propionyl CoA to methylmalonyl CoA
E. methylmalonyl CoA to succinyl CoA.

12. The highest amount of ATP per gram is yielded by metabolism of:

A. sucrose
B. starch
C. glutamic acid
D. oleic acid
E. succinic acid.

13. 1,25-Dihydroxycholecalciferol (calcitriol) is the active form of:

A. vitamin A
B. vitamin B
C. vitamin C
D. vitamin D
E. vitamin E.

14. In the mammal, tryptophan is a precursor of all of the following EXCEPT:

A. NAD
B. Nicotinate
C. NADP
D. FAD
E. serotonin.

15. The four D's of dementia, diarrhea, dermatitis, and death are associated with a deficiency of:

A. vitamin A
B. vitamin B_{12}
C. vitamin C
D. biotin
E. niacin.

16. Which of the following has a quinone structure related to that of vitamin K?

A. Coenzyme Q
B. histidine
C. thyroxine
D. vitamin D
E. vitamin E.

17. Which of the following cofactors does NOT require a specific dietary vitamin precursor?

A. FAD
B. NAD
C. ATP
D. Coenzyme A
E. ACP.

18. You have just consumed a meal containing 200g carbohydrate, 40g protein, 40g fat, and 30g ethanol. How many kilocalories did you take in?

A. 960
B. 1040
C. 1280
D. 1400
E. 1530.

19. Parathyroid hormone increases the blood level of:

A. Mg^{++}
B. Cu^{++}
C. Ca^{++}
D. Fe^{++}
E. Na^+.

20. Wernicke's disease may follow a deficiency of:

A. thiamin
B. riboflavin
C. niacin
D. folic acid
E. cyanocobalamin.

21. Starvation results in:

A. decrease in liver phospholipid
B. increase in urinary acetoacetic acid
C. early breakdown of muscle protein
D. decrease in gluconeogenesis
E. higher levels of insulin.

22. A derivative of folic acid is coenzyme for all of the following EXCEPT:

A. formation of thymidylate
B. formation of serine from glycine
C. formation of purines
D. metabolism of 1-C units
E. formation of urea.

23. Utilization of propionate involves:

A. B_{12}
B. Pantothenic acid
C. biotin
D. riboflavin
E. all the above.

24. Oxygenases which replace C-H by C-OH require:

A. NADH
B. tetrahydrofolic acid
C. cytochrome P-450
D. H_2O_2
E. superoxide.

25. Human dietary requirements include:

A. cysteine
B. cytosine or uracil
C. linoleic acid
D. glutamic acid
E. tyrosine.

26. Raw egg white contains a protein, avidin, which:

A. digests fibrin clots
B. acts like a trypsin inhibitor
C. produces a protein deficiency when fed
D. prevents absorption of biotin
E. blocks the formation of choline from betaine.

MATCHING: For each set of questions, choose the ONE BEST answer from the lettered list above it. An answer may be used one or more times, or not at all.

Questions 27-30:

    A. transaldolase
    B. methylmalonyl CoA mutase
    C. α-ketoglutarate decarboxylase
    D. cytochrome oxidase
    E. acetyl CoA carboxylase

27. Biotin.

28. Adenosylcobalamin.

29. Iron and copper.

30. Thiamin.

Questions 31-34: A deficiency of

    A. vitamin A
    B. zinc
    C. iron
    D. folate
    E. iodide

will cause:

31. Megaloblastic anemia.

32. Hypochromic anemia.

33. Endemic goiter.

34. Xerophthalmia.

Questions 35-39: A cofactor derived from

    A. thiamin
    B. pantothenic acid
    C. nicotinic acid
    D. pyridoxine
    E. biotin

is involved in:

35. Carboxylation of propionate.

36. Transamination of amino acids.

37. Transketolase.

38. Activation of palmitic acid.

39. Oxidation of lactate.

Questions 40-43:

    A. cobalt
    B. molybdenum and copper
    C. copper only
    D. manganese
    E. zinc

40. Carbonic anhydrase.

41. Vitamin B_{12}.

42. Ferroxidase.

43. Xanthine oxidase.

Questions 44 - 47:

    A. pyridoxal phosphate
    B. biotin
    C. thiamin pyrophosphate
    D. vitamin B_{12}
    E. tetrahydrofolic acid

44. Transaminase.

45. Pernicious anemia.

46. Pyruvate carboxylase.

47. transformylase.

Questions 48 - 51: The following pathways:

    A. tricarboxylic acid cycle starting from acetyl CoA
    B. glycogen synthesis from glucose
    C. gluconeogenesis starting from pyruvate
    D. glycolysis leading to lactate
    E. pentose phosphate pathway

require cofactors derived from the vitamins:

48. Thiamin and nicotinamide only.

49. Nicotinamide only.

50. Nicotinamide, thiamin, riboflavin, and pantothenic acid.

51. Nicotinamide and biotin.

Questions 52 - 57: A deficiency of:

    A. ascorbic acid
    B. nicotinic acid
    C. thiamin
    D. vitamin D
    E. vitamin K

results in:

52. Pellagra.

53. Beriberi.

54. Hemorrhagic disease of the newborn.

55. Scurvy.

56. Hypoprothrombinemia.

57. Rickets.

V. ANSWERS TO QUESTIONS ON NUTRITION

1.	C	21.	B	41.	A
2.	D	22.	E	42.	C
3.	E	23.	E	43.	B
4.	B	24.	C	44.	A
5.	B	25.	C	45.	D
6.	C	26.	D	46.	B
7.	E	27.	E	47.	E
8.	E	28.	B	48.	E
9.	C	29.	D	49.	D
10.	E	30.	C	50.	A
11.	E	31.	D	51.	C
12.	D	32.	C	52.	B
13.	D	33.	E	53.	C
14.	D	34.	A	54.	E
15.	E	35.	E	55.	A
16.	A	36.	D	56.	E
17.	C	37.	A	57.	D
18.	E	38.	B		
19.	C	39.	C		
20.	A	40.	E		

11. PURINES AND PYRIMIDINES

Leon Unger

Nucleic acids are comprised of nitrogenous bases (purines and pyrimidines), pentose sugars (ribose and deoxyribose) and phosphate groups. Specific sequences of purines and pyrimidines encode the genetic information of cells and organisms.

I. STRUCTURE AND NOMENCLATURE

A. Nitrogenous Bases

The purines, **adenine (A)** and **guanine (G)**, are present in both RNA and DNA. Catabolism of adenine and guanine produces the purines **inosine, hypoxanthine, xanthine** and **uric acid.**

The pyrimidines, **cytosine (C)** and **thymine (T)** are present in DNA. Cytosine and **uracil (U)** are contained in RNA (Figure 11-1).

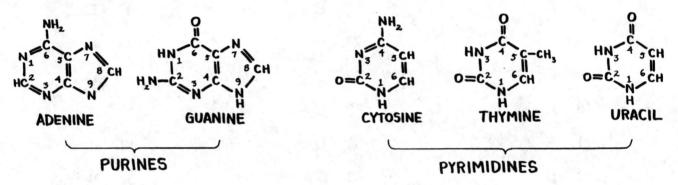

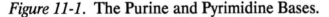

Figure 11-1. The Purine and Pyrimidine Bases.

B. Nucleosides

A **nucleoside** consists of a purine or pyrimidine base linked to a pentose sugar. In RNA, the pentose is **D-ribose**, therefore the nucleoside formed is termed a <u>ribonucleoside</u> (example: adenosine). The pentose is **2-deoxy-D-ribose** in DNA, and the nucleoside is called a <u>deoxyribonucleoside</u> (example: deoxyadenosine). The atoms of the pentoses are designated by primed numbers to distinguish them from the numbers in the bases (Figure 11-2).

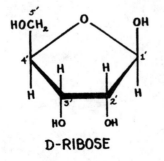

D-RIBOSE

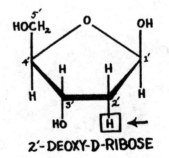

2'-DEOXY-D-RIBOSE

Figure 11-2. The Pentoses of Nucleic Acids.

In a nucleoside, carbon 1′ (C-1′) of the pentose is linked by a β-glycosidic bond to N-9 of the purine or N-1 of the pyrimidine.

N-9-purine   N-1-pyrimidine
| |
pentose-C-1′   pentose-C-1′

The major ribonucleosides for the purines are **adenosine** and **guanosine** and for the pyrimidines, **cytidine** and **uridine**.

The major deoxyribonucleosides are **deoxyadenosine, deoxyguanosine, deoxycytidine** and (deoxy)-**thymidine** (Figure 11-3).

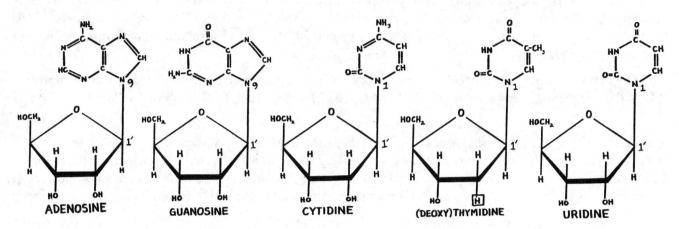

Figure 11-3. The Major Nucleosides.

C. Nucleotides

A nucleotide is a nucleoside in which a phosphate group has been esterified to a hydroxyl group on the pentose. The most common position to be esterified is at C-5′. Other positions for esterification include the hydroxyls at C-3′ (common) and C-2′ (less common).

II. SYNTHESIS OF NUCLEOSIDE DIPHOSPHATES AND TRIPHOSPHATES

The ribo- or deoxyribo-nucleoside 5′-monophosphates [NMP, (d)NMP] may be further phosphorylated. Using ATP as a phosphate donor, *nucleoside monophosphate kinase* phosphorylates (d)NMP to (d)NDP. *Nucleoside diphosphate kinase* then converts (d)NDP to (d)NTP. These nucleotides are readily interconvertible.

(d)NMP + ATP ⇌ (d)NDP + ADP

(d)NDP + ATP ⇌ (d)NTP + ADP

5-Phosphoribosyl-1-Pyrophosphate (PRPP)

Functions: PRPP supplies ribose phosphate for the synthesis of <u>purine</u> ribonucleotides by both the *de novo* pathway and the salvage pathway. It also supplies ribose phosphate for the synthesis of <u>pyrimidine</u> ribonucleotides. PRPP is an intermediate in **histidine** and **tryptophan** biosynthesis.

Synthesis: PRPP is synthesized from ribose-5-phosphate and ATP. The enzyme catalyzing this reaction is *ribose phosphate pyrophosphokinase (phosphoribosylpyrophosphate synthetase).*

$$\text{Ribose-5-P} + \text{ATP} \longrightarrow \text{PRPP} + \text{AMP}$$

III. PURINE METABOLISM

A. Biosynthesis by the *De Novo* Pathway

The atoms for the purine ring are derived from **amino acids, tetrahydrofolate** and CO_2 and are assembled stepwise onto a pre-existing ribose phosphate molecule. The amino acids which contribute C and N atoms are **glycine, aspartic acid** and **glutamine.** Glycine supplies C-4, C-5 and N-7. Aspartate provides N-1. Glutamine supplies N-3 and N-9 from the amido group. The ribose phosphate comes from the **hexose monophosphate pathway.**

The first nucleotide synthesized is **inosinate (IMP).** The purine base in this nucleotide is **hypoxanthine.** IMP is then sequentially convertible to AMP and GMP. These are then further phosphorylated to ATP and GTP.

The first step specific to the synthesis of IMP is the formation of **5-phosphoribosylamine** from PRPP and glutamine. In this reaction, mediated by *amidophosphoribosyl transferase*, the pyrophosphate group of PRPP is replaced by an amino group donated by glutamine. The irreversible formation of phosphoribosylamine commits PRPP to purine synthesis. The amidotransferase is inhibited by **azaserine.**

Glutamine   Glutamate

PRPP $\longrightarrow$ 5-phosphoribosylamine + PP_i

The next step involves the addition of the entire **glycine** molecule. The glycine atoms become C-4, C-5 and N-7 of the purine ring. This step requires the utilization of one ATP.

C-8 is provided by the methenyl group of N^5, N^{10}-**methenyltetrahydrofolate.**

N-3 is derived from the amide group of **glutamine.** This step utilizes an ATP.

The next step involves the cyclization of the molecule to form the five-membered **imidazole** ring.

C-6 is added by carboxylation with CO_2. **Biotin** is a cofactor in this process.

N-1 is provided by the amino group of **aspartate.** Consumption of a molecule of ATP is required.

C-2 is derived from the formyl group of N^{10}-**formyltetrahydrofolic acid.**

Then ring closure to form the first complete purine, IMP, or hypoxanthine ribose phosphate:

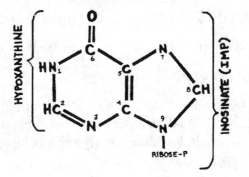

The origins of the atoms in the purine ring are summarized in Figure 11-4.

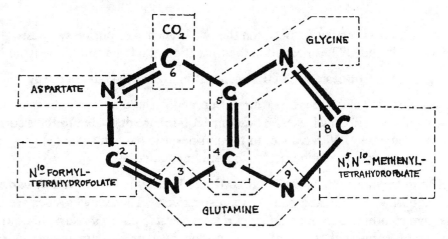

Figure 11-4. Origin of the Atoms in Purines.

Finally, AMP and GMP are formed from IMP as shown in Figure 11-5.

B. Control of *De Novo* Purine Nucleotide Biosynthesis

There are three sites at which feedback inhibition regulates *de novo* purine nucleotide biosynthesis (Figure 11-5):

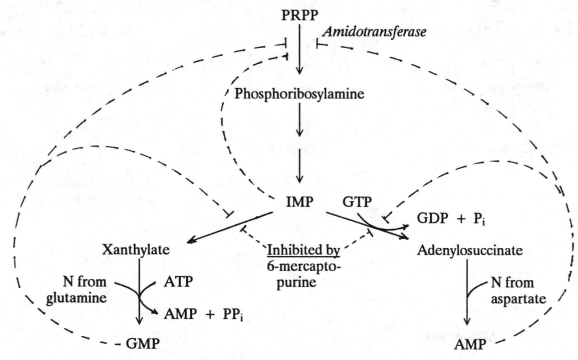

Figure 11-5. Synthesis of Adenylate and Guanylate, and Regulation by Nucleotides.

1. Excess AMP and GMP inhibit the conversion of PRPP to phosphoribosylamine. These end-products synergistically inhibit *amidophosphoribosyl transferase*, the first enzyme specific for purine biosynthesis.

2. AMP inhibits the reaction converting IMP to **adenylosuccinate**, thereby preventing further AMP synthesis. This inhibition does not affect GMP formation.

3. GMP inhibits the conversion of IMP to **xanthylate** (XMP), preventing its own formation without affecting AMP synthesis.

In addition, there is a cross-regulation between the two branches of purine synthesis: ATP is required to convert xanthylate to GMP, and GTP is needed in the formation of adenylosuccinate from IMP.

C. Catabolism of Purine Nucleotides and Their Regeneration by the Salvage Pathway

During digestion, nucleic acids are cleaved to purine and pyrimidine nucleotides by the action of pancreatic ribonucleases and deoxyribonucleases. These nucleotides are degraded to their corresponding nucleosides and then to the free bases. With respect to the purines, the bases may be (1) <u>catabolized</u> to urate (in man, Figure 11-6) or (2) <u>salvaged</u> to regenerate the parent nucleotides.

Catabolism. The nucleotides AMP, IMP and GMP are hydrolytically cleaved by *purine 5′-nucleotidase* to the nucleosides adenosine, inosine and guanosine, respectively. These nucleosides are catabolized to the free bases adenine, hypoxanthine and guanine by *purine nucleoside phosphorylase*, liberating ribose-1-phosphate. Instead of being converted to adenine, adenosine may be deaminated to inosine by *adenosine deaminase*.

If hypoxanthine is not processed through the salvage pathway and recycled, it is oxidized to **xanthine** and then to **uric acid**; these two final steps are both catalyzed by *xanthine oxidase*. Guanine may be deaminated to xanthine by *guanine deaminase*, then converted to uric acid. **Uric acid** is the final principal end-product of purine catabolism in man and is excreted in the urine (Figure 11-6).

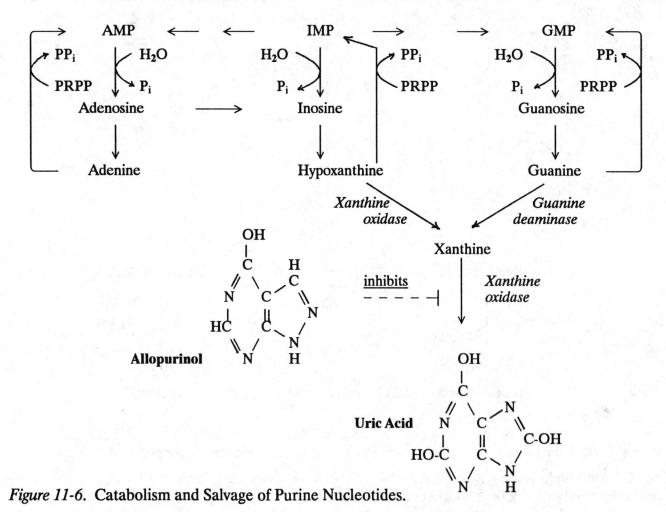

Figure 11-6. Catabolism and Salvage of Purine Nucleotides.

Salvage. One salvage enzyme, *adenine phosphoribosyl transferase (APRTase)*, catalyzes the transfer of ribosyl phosphate from PRPP to adenine to resynthesize AMP. A second salvage enzyme, *hypoxanthine-guanine phosphoribosyl transferase (HGPRTase)*, mediates the transfer of ribose phosphate to hypoxanthine or guanine to regenerate IMP or GMP, respectively:

$$\text{Adenine} + \text{PRPP} \xrightarrow{(1)} \text{AMP} + \text{PP}_i$$

$$\text{Hypoxanthine} + \text{PRPP} \xrightarrow{(2)} \text{IMP} + \text{PP}_i$$

$$\text{Guanine} + \text{PRPP} \xrightarrow{(2)} \text{GMP} + \text{PP}_i$$

1 = *adenine phosphoribosyl transferase (APRTase)*

2 = *hypoxanthine-guanine phosphoribosyl transferase (HGPRTase)*

D. Disorders of Purine Metabolism

1. *Gout* is the result of the excessive production of uric acid and its subsequent deposition, as poorly soluble sodium urate crystals, in the joints and kidneys. The overproduction of urate results from the catabolism of abnormally high levels of purines synthesized via the *de novo* pathway. The high concentration of purines may be due to an increase in PRPP either because of (1) decreased utilization of PRPP by the salvage pathway caused by a partial deficiency of *HGPRTase* activity, or (2) <u>enhanced production</u> of PRPP because of overactivity of *phosphoribosyl pyrophosphate synthetase*. High levels of purines may also result from an increase in **phosphoribosylamine (PR-amine)** due to overactivity of *amidophosphoribosyl transferase* which is the rate-limiting enzyme in purine synthesis.

Gout is treated with **allopurinol**, an analog of hypoxanthine (Figure 11-6). Allopurinol inhibits *xanthine oxidase* and, therefore, prevents the conversion of hypoxanthine to xanthine and then xanthine to urate. This results in a reduction of serum urate concentration and increases the concentrations of the more soluble hypoxanthine and xanthine.

2. The *Lesch-Nyhan Syndrome* results from a nearly complete absence of *HGPRTase* because of a genetic defect. Consequently, there is an absence or marked decrease of hypoxanthine and guanine salvage, an elevated PRPP concentration and an increased purine nucleotide synthesis by the *de novo* pathway. The ensuing catabolism of the over-produced purines leads to increased urate production. Victims of this disease show <u>gout</u>, <u>mental retardation</u>, <u>self-mutilation</u>, <u>hostility</u> and a <u>lack of muscular coordination</u>. Allopurinol is not effective in alleviating the symptoms of this disease. The syndrome is inherited as a <u>sex-linked recessive</u> trait in males.

3. *Immunodeficiency Diseases*

a. A deficiency of *adenosine deaminase* is associated with a severe combined deficiency of both T-cells and B-cells and deoxyadenosinuria.

b. A *purine nucleoside phosphorylase* deficiency leads to a severe T-cell deficiency but normal B-cell function. Additional characteristics of this disorder include inosinuria, guanosinuria and hypouricemia. Both of these disorders are inherited as <u>autosomal recessive</u> traits.

IV. PYRIMIDINE METABOLISM: BIOSYNTHESIS

Whereas the growing purine ring is assembled on ribose-P, the pyrimidine ring is formed first and then joined to ribose-P to form a pyrimidine nucleotide. (The steps below are identified in Figure 11-7).

Step 1: The regulatory step in the synthesis of pyrimidines in mammals is the formation of **carbamoyl phosphate** by cytoplasmic *carbamoyl phosphate synthetase*. This enzyme is inhibited by UTP, a product of the pathway.

Step 2: The committed step in the synthesis of the pyrimidine ring in bacteria is the formation of **N-car-bamoylaspartate** from carbamoyl phosphate and aspartate. The carbamoyl phosphate atoms become C-2 and N-3. Aspartate donates C-4, C-5, C-6 and N-1. The carbamoylation of aspartate is catalyzed by *aspartate transcarbamoylase (ATC-ase)*. In bacteria, ATC-ase is the rate limiting enzyme; it is feedback-inhibited al-losterically by CTP, the end-product of the pathway.

Steps 3 and 4: Ring closure and oxidation yields the pyrimidine, **orotic acid**.

Step 5: Ribose-P, derived from PRPP, is attached to orotate to yield the nucleotide, **orotidylate (orotidine-5-P)**.

Step 6: Orotidylate is decarboxylated to form **UMP**. All other pyrimidine nucleotides are synthesized from UMP.

Steps 7 and 8: UMP is phosphorylated sequentially to **UDP** and **UTP**.

Step 9: UTP is aminated by glutamine to form **CTP**.

V. DEOXYRIBONUCLEOTIDES ARE FORMED BY REDUCTION OF RIBONUCLEOSIDE DIPHOSPHATES.

Deoxyribonucleotides are formed from ribonucleotides. The ribose of the ribonucleoside diphosphate is reduced at the 2′ carbon atom to form the 2′-deoxyribonucleoside diphosphate.

$$NDP \longrightarrow dNDP$$

$$\text{For example, UDP} \longrightarrow dUDP$$

DNA also differs from RNA in that it contains **thymine** instead of uracil. Thymine is uracil methylated at the C-5 position. **Deoxythymidylate (dTMP)** is produced by the methylation of deoxyuridylate (dUMP) by tetrahydrofolate (Figure 11-7).

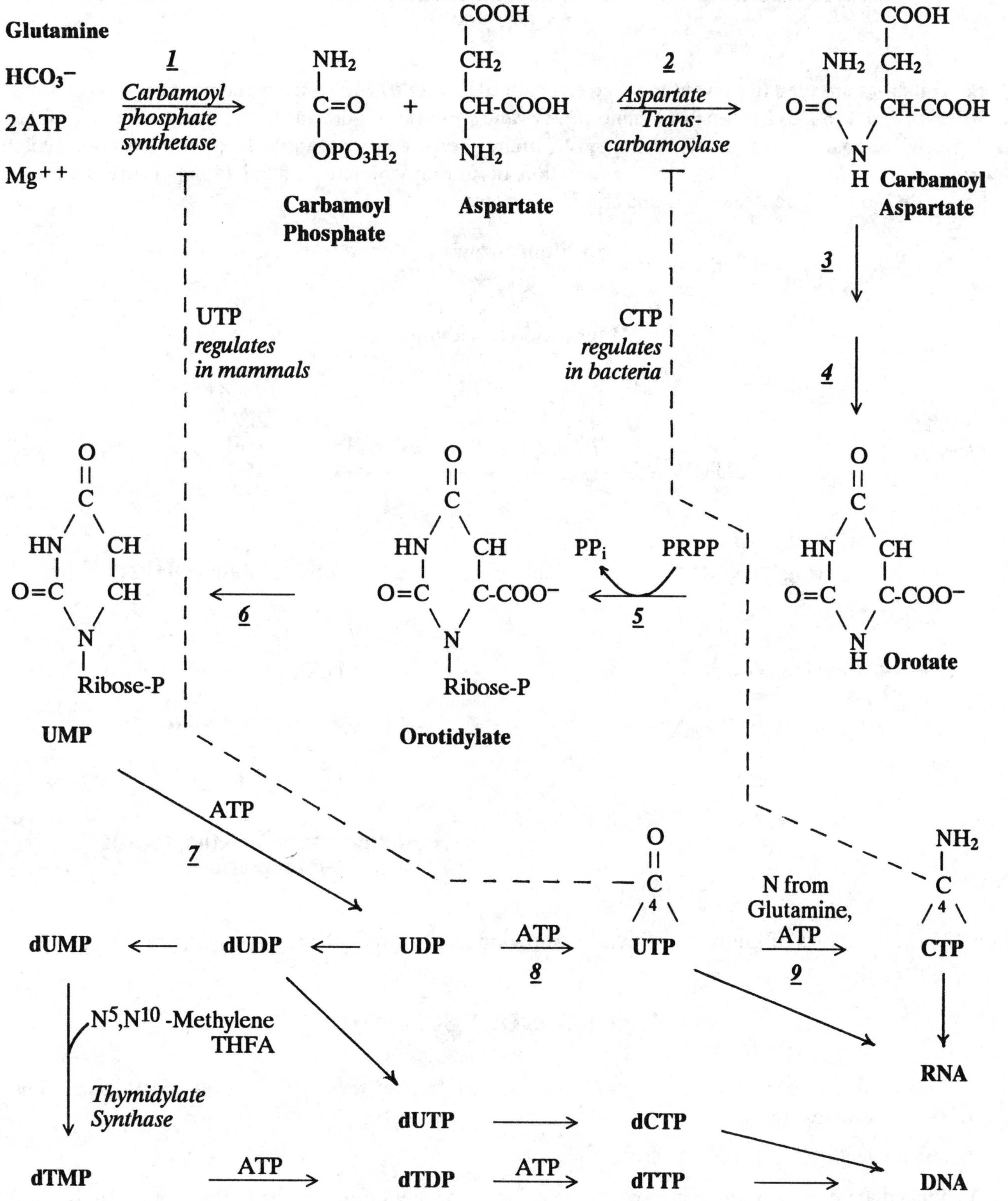

Figure 11-7. Biosynthesis of Pyrimidine Nucleotides, and Sites of Feedback Control. Underlined numbers refer to steps described in text.

VI. SOME ANTICANCER DRUGS ACT BY BLOCKING DEOXYTHYMIDYLATE SYNTHESIS

Several drugs are used to prevent the conversion of dUMP to dTMP which is required for DNA synthesis. **5-Fluorouracil (5-FU)** irreversibly inhibits *thymidylate synthase* (Figure 11-8) and is particularly useful in treating solid tumors. The folic acid antagonists, **aminopterin** and **amethopterin (methotrexate)**, inhibit *dihydrofolate reductase*, preventing the regeneration of tetrahydrofolate and inhibiting dTMP synthesis. Methotrexate is used to treat leukemia and choriocarcinoma.

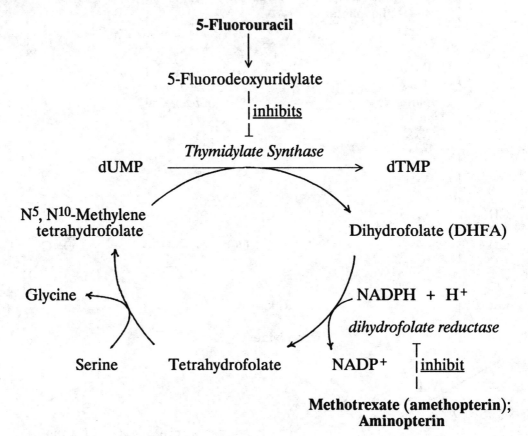

Figure 11-8. Synthesis of Deoxythymidylate, Showing Inhibition by Some Anti-Cancer Drugs.

VII. FUNCTIONS OF NUCLEOTIDES

The nucleoside diphosphates and triphosphates are the "active" forms of the nucleotides. Some functions of the nucleotides are:

A. Adenine nucleotides

1. Immediate energy source for most enzymatic reactions requiring energy expenditure. The form in which most biological energy is stored.

2. Precursor to RNA and DNA.

3. Precursor to cAMP.

4. Precursor to important coenzymes like NAD^+, FAD and CoA.

5. Regulator of many enzymes by phosphorylation of serine or threonine or tyrosine residues. In some instances, an adenosyl group is transferred to the enzyme.

6. Regulator of many enzymes by allosteric mechanisms; effector may be in the form of ATP, ADP or AMP.

B. Guanosine nucleotides

1. Precursor to RNA and DNA.

2. Precursor to cGMP.

3. Nucleotide carrier for mannose and fucose in glycoprotein biosynthesis.

4. Involved in several key steps in peptide bond formation in protein biosynthesis.

5. Key nucleotide in biochemistry of vision.

C. Uridine nucleotides

1. Precursor to RNA and DNA.

2. Nucleotide carrier for glucose, galactose, N-acetylglucosamine and glucuronic acids in polysaccharide, glycoprotein and glycosaminoglycan biosyntheses.

D. Cytidine nucleotides

1. Precursor to RNA and DNA

2. Nucleotide carrier for diacylglycerol in phosphoglyceride synthesis and choline salvage.

3. Nucleotide carrier for neuraminic acid (sialic acids) in glycoprotein synthesis.

VIII. REVIEW QUESTIONS ON PURINES AND PYRIMIDINES

DIRECTIONS: For each of the following multiple-choice questions (1 - 14), choose the ONE BEST answer.

1. The four nitrogen atoms of the purine ring are derived from:

A. aspartate, glutamine, and glycine
B. glutamine, ammonia, and aspartate
C. glycine and aspartate
D. ammonia, glycine, and glutamate
E. urea and ammonia

2. Thymidylate synthase catalyzes the conversion of:

A. dUDP → dUMP
B. dUMP → dTMP
C. dTMP → dTDP
D. dTDP → dTTP
E. dCTP → dUTP

3. By which of the following means are bases linked to the pentoses in RNA and DNA molecules?

A. They are linked together by N-glycosidic bonds in the beta-configuration
B. They are linked together by $3',5'$ phosphodiester bridges
C. They are linked together by $2',5'$ phosphodiester bridges
D. They are linked together by alpha-glycosidic bonds
E. They are held together by electrostatic attraction

4. Thymine is:

A. 2,4-dioxy-pyrimidine
B. 6-aminopurine
C. 2-amino-6-oxy-purine
D. 2,4-dioxy-5-methyl-pyrimidine (5-methyl uracil)
E. 2-oxy-4-amino-pyrimidine

5. Deoxyribonucleotides are formed by reduction of:

A. ribonucleosides
B. ribonucleoside monophosphates
C. ribonucleoside diphosphates
D. ribonucleoside triphosphates
E. deoxyribonucleoside triphosphates

6. The oxidation of xanthine to urate is catalyzed by:

A. aspartic transcarbamoylase
B. xanthine oxidase
C. hypoxanthine-guanine phosphoribosyl transferase
D. purine nucleoside phosphorylase
E. purine $5'$-nucleotidase

7. AMP is synthesized in a two step reaction sequence involving

A. inosinic acid, NAD^+ and glutamine
B. inosinic acid, ATP and glutamine
C. inosinic acid, GTP and aspartate
D. hypoxanthine and ribose-1-phosphate
E. none of these

8. An inherited immunodeficiency disease in which there is a very severe T-cell defect, but normal B-cell function, is associated with a deficiency of an enzyme of purine catabolism. The deficient enzyme is:

A. aspartic transcarbamoylase
B. xanthine oxidase
C. hypoxanthine-guanine phosphoribosyl transferase
D. purine nucleoside phosphorylase
E. purine $5'$-nucleotidase

9. Each of the following compounds is used in the biosynthesis of BOTH purine and pyrimidine nucleotides EXCEPT:

A. glutamine
B. aspartic acid
C. 5-phosphoribosyl-1-pyrophosphate
D. carbamoyl phosphate
E. tetrahydrofolic acid derivative

10. The committed step in purine synthesis is the step in which:

A. phosphoribosylamine is synthesized from PRPP and glutamic acid
B. glutamine is incorporated intact into the molecule on the sugar phosphate unit
C. N-formylglycinamide ribonucleotide is formed from glycinamide ribonucleotide with a methyl group transferred from N^5, N^{10}-methylene THFA
D. phosphoribosylamine is formed enzymatically from PRPP and glutamine
E. bicarbonate is utilized to carboxylate aminoimidazole ribonucleotide

11. In the purine catabolic pathway, adenosine is:

A. oxidized directly to xanthine
B. converted to guanine and then deaminated to xanthine
C. deribosylated to adenine or deaminated to inosine
D. phosphorylated to give IMP
E. oxidized directly to urate

12. Important controls of purine nucleotide *de novo* synthesis include:

A. availability of PRPP
B. inhibition by GMP and AMP of glutamine-PRPP amidotransferase
C. GTP requirement for AMP synthesis
D. ATP requirement for GMP synthesis
E. all of the above

13. Gout is a disorder frequently associated with a partial deficiency of an enzyme of purine metabolism. The deficient enzyme is:

A. adenosine deaminase
B. adenine phosphoribosyl transferase
C. xanthine oxidase
D. hypoxanthine-guanine phosphoribosyl transferase
E. purine nucleoside phosphorylase

14. The committed step in pyrimidine biosynthesis in *E. coli* involves which of the following substrates?

A. IMP and aspartic acid
B. orotic acid and 5-phosphoribosyl-1-pyrophosphate (PRPP)
C. aspartic acid and ribose-5-phosphate
D. aspartic acid and carbamoyl phosphate
E. PRPP and phosphoribosylamine

MATCHING: For each set of questions, choose the ONE BEST answer from the list of lettered options above it. An answer may be used one or more times, or not at all.

Questions 15 - 17:

    A. Severe combined (T cell and B cell) immunodeficiency disease
    B. Lesch-Nyhan syndrome
    C. T cell immunodeficiency disease
    D. Beta-thalassemia
    E. I-disease

15. Hypoxanthine-guanine phosphoribosyl transferase (HGPRT) deficiency.

16. Purine nucleoside phosphorylase deficiency.

17. Adenosine deaminase (ADA) deficiency.

IX. ANSWERS TO QUESTIONS ON PURINES AND PYRIMIDINES

1.	A		10.	D
2.	B		11.	C
3.	A		12.	E
4.	D		13.	D
5.	C		14.	D
6.	B		15.	B
7.	C		16.	C
8.	D		17.	A
9.	D			

12. NUCLEIC ACIDS: STRUCTURE AND SYNTHESIS

Jay Hanas

I. DNA

A. Structure

Deoxyribonucleic acid (DNA) and ribonucleic acid (RNA) are responsible for the transmission of genetic information and are intimately involved in cellular metabolism, growth, and differentiation. The central dogma of Molecular Biology states that genetic information is transferred from DNA to RNA to protein.

$$DNA \longrightarrow RNA \longrightarrow Protein$$

The understanding of the structure, synthesis, and function of DNA and RNA is of fundamental importance in biology and medicine. DNA is a polymer of deoxynucleoside monophosphates (deoxynucleotides). The $5'$ phosphate group of one deoxynucleotide is joined to the $3'$ OH group of another, forming a phosphodiester linkage. A polymer of nucleotides (polynucleotide) results when many nucleotides are joined in linear fashion. In deoxyribonucleic acid (DNA, lacking a $2'$ OH group on the nucleotide sugar), the nucleotide bases (adenine, guanine, cytosine, and thymine) of one strand form hydrogen bonds with the nucleotide bases of the other strand (Figure 12 - 1).

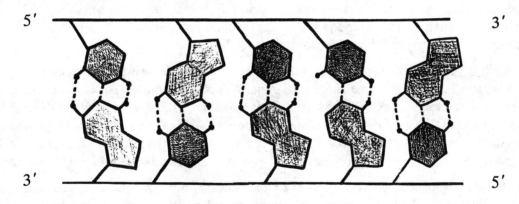

Figure 12 - 1. Pairing of Bases in Double-stranded DNA.

There are precise rules governing which bases form hydrogen bonds with one another: adenine forms two hydrogen bonds with thymine; guanine forms three hydrogen bonds with cytosine. When DNA forms a double stranded structure, the polynucleotides are anti-parallel, i.e. one strand is in the $5'$-$3'$ orientation and the other strand is in the opposite $3'$-$5'$ orientation. DNA can also exist in a single-stranded form. DNA strands can contain millions of nucleotides joined to each other by phosphodiester linkages. Genetic information is stored as unique sequences of the four nucleotide bases. The **genome** of an organism is all of the DNA content expressed as base pairs. The human genome is much larger than those of bacteria and unicellular organisms (by several orders of magnitude) but smaller than that of some plants and amphibians. The significance of the differences in genome sizes is not known, but in general more complex organisms have larger genomes. Usually, more complex organisms also have large amounts of repetitive DNA which does not appear to code for functional proteins. The function of this repetitive DNA is unknown.

Double-stranded DNA is the repository of genetic information and in animal cells is located in the nucleus and mitochondria. The DNA in the nucleus is very large in size (millions of base pairs) whereas the DNA in animal mitochondria is small by comparison and exists in a circular form (the ends are covalently joined). Mitochondrial DNA codes for some proteins involved in electron transport processes and for other molecules necessary for the functions of these organelles.

Double-stranded DNA forms a double helix structure in which the two strands are intertwined as two right-handed helices (the helices in DNA turn clockwise as one looks up the center of the molecule). One complete turn of the double helix requires 10.4 base pairs in the "B" form of DNA. The two strands of a duplex DNA molecule unwind (breaking of hydrogen bonds) in a process called **denaturation**. Denaturation can be brought about by high temperature, alkali, or specific enzymes. The T_m (melting point) of DNA is the temperature at which 50% of the bases become unpaired and is proportional to the GC content of the molecule. The individual strands can re-anneal in a process called **renaturation** if the DNA is cooled, or the pH brought to neutrality.

The ability of complementary DNA sequences to re-anneal or **hybridize** is the basis for identifying specific DNA sequences of clinical interest. For example, the presence of a particular virus in blood can be detected with a radiolabeled, single-stranded DNA **probe** which hybridizes only to the viral genome and not to lymphocyte DNA.

The individual nucleotide bases in double-stranded nucleic acids absorb less UV light than in single-stranded or free forms, a phenomenon called **hypochromicity**. When DNA is denatured, UV absorbance of the sample increases (hyperchromicity). Single-stranded nucleic acids can form intrastrand base-pairs giving partial double stranded character (hairpin structures).

The genomes of bacteria, some viruses, mitochondria and plasmid DNA are covalently closed, circular DNA molecules. Circular, double-stranded DNA is topologically restrained (as is DNA in which both ends are anchored by some means) and in many cases is **supercoiled**.

Supercoiling is governed by enzymes termed **topoisomerases**. There are basically two types of DNA topoisomerases, type I or type II. Type II enzymes require ATP as an energy source whereas type I enzymes require no exogenous energy source. Topo I in bacteria relaxes negatively supercoiled DNA whereas topo II in bacteria (also called DNA gyrase) promotes negative supercoiling. *In vivo*, the circular genome of bacteria and plasmids exists in a negative supercoiled state. Negative supercoiling makes DNA melting during replication or transcription energetically favorable. DNA gyrase is inhibited by **nalidixic acid** and **novobiocin**.

Eukaryotic organisms contain type I and II topoisomerases but not a DNA gyrase, the enzyme which promotes negative supercoiling. Negative supercoiling in eukaryotic organisms can be generated through the formation and removal of nucleosomes in which the DNA is wrapped around a nucleosome about one and three-quarters times.

B. DNA Replication

DNA replication results in the exact duplication of the double-stranded DNA molecule. DNA replication takes place by a semiconservative pathway in which the two strands of duplex DNA are separated and each single strand serves as a template for the synthesis of the second, complementary strand (Figure 12 - 2).

DNA replication in bacteria is better understood than in mammalian cells. The bacterium *E. coli* contains three distinct DNA polymerases (the enzyme which catalyzes the polymerization of deoxynucleotides on the template strand), called DNA pol I, II, and III. Mammalian cells also contain three DNA polymerases, two nuclear (one for repair) and a distinct mitochondrial enzyme. DNA polymerases require a free 3′ OH group in order to react with an incoming deoxynucleotide 5′ triphosphate. Thus, the template strand needs a hybridized primer to initiate DNA synthesis. The primer is usually a small RNA strand (RNA polymerase does not need a primer to begin synthesis). DNA pol I in bacteria is the most abundant DNA polymerase and is used for repair synthesis of DNA.

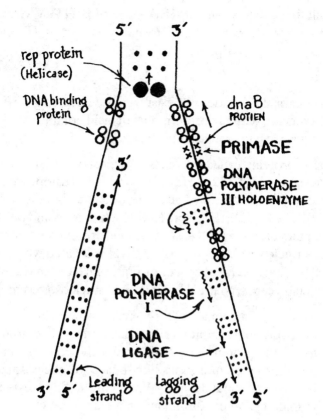

Figure 12 - 2. DNA Replication in Prokaryotes.

DNA synthesis is continuous on one strand (**leading strand**) and discontinuous on other (**lagging strand**). This means one strand of the parental DNA duplex is copied at the replication fork by DNA polymerase III in a continuous fashion whereas the other strand needs a "back-stitching" process to synthesize the other strand. Primers are of continual necessity for this back-stitching process (Figure 12 - 2). The discontinuous nature of DNA synthesis results in the production of newly synthesized DNA pieces called **Okazaki fragments**. DNA ligase joins the Okazaki fragments once the RNA primer has been degraded and the gaps filled in. DNA polymerase I also can correct errors in synthesis because it has a 3′ to 5′ exonuclease activity which is capable of cleaving off nucleotides which were mis-incorporated.

DNA replication is bi-directional, i.e. two replication forks (where the duplex DNA is denatured into two single strands), initiated at the same point, move in opposite directions on the same duplex strand. In bacteria there is usually only one origin of replication but in mammalian cells, there are thousands.

DNA replication requires the unwinding of the duplex DNA template to yield the single-stranded templates. Energy is needed to unwind or denature duplex DNA. Cells contain enzymes called **helicases** which can unwind the duplex DNA molecule utilizing ATP as an energy source. This unwinding process associated with DNA replication would eventually come to a halt because of topological constraints due to the build up of "knots" ahead of the replication fork. Topoisomerases relax these knots (by nicking and rejoining the DNA) and thus permit the replication fork to proceed.

A number of medically important animal viruses, e.g. HTLV (human T-cell lymphotropic virus) utilize an enzyme called **reverse transcriptase** to convert their single-stranded RNA genome into a duplex DNA genome which integrates into nuclear DNA. This is another example of DNA replication although in this case the template molecule is a single-stranded RNA molecule. HTLV viruses cause some forms of human leukemia. HIV (human immunodeficiency virus, formerly known as HTLV-III) is the causative agent in AIDS (acquired immune deficiency syndrome).

C. Deoxyribonucleoproteins

In most cases, DNA in animal cells does not exist free but occurs tightly bound to specific proteins. This is because the functional units are actually protein-nucleic acid complexes. In addition, the protein components protect the nucleic acids from degradation by nucleases.

The proteins are basic proteins called **histones** which contain large amounts of **lysine** and **arginine**. There are five histones, designated H1, H2A, H2B, H3, and H4. Histones are highly conserved proteins with H3 and H4 being the most highly conserved of all proteins. They interact with each other in a very ordered way first to form dimers of H2A, H2B, H3, and H4, which then combine with each other to produce an **octomer**. About 145 base pairs of duplex DNA are wrapped around an octomer (slightly less than two turns) to form a structure called a **nucleosome** (Figure 12 - 3). H1 is located where DNA enters and leaves the nucleosome. About 25 - 100 bp (base pairs) of spacer or linker DNA span the region between adjacent nucleosomes. A single DNA molecule can wrap itself around many nucleosomes to form a **polynucleosome**.

Chromatin and Chromosomes. In animal cells, polynucleosomes can be packaged along with many non-histone proteins into a nucleoprotein complex called **chromatin**. If chromatin is washed extensively with concentrated salt solutions, all the proteins come off except the core histones, H2A, H2B, H3, H4. In the electron microscope the remaining structure looks like "beads" of nucleosomes on a "string" of DNA. Each "bead on a string" has a diameter of about 10 nm. When chromatin is less extensively washed with salt, the observed structure has a diameter of about 30 nm and appears to be composed of a helical array of nucleosomes (Figure 12 - 3).

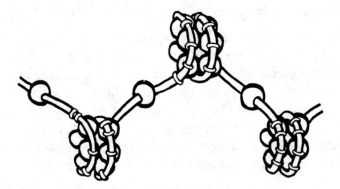

Figure 12 - 3. Interactions between DNA and Histones to Form Nucleosomes.

Chromosomes are highly condensed structures consisting of the 30-nm chromatin fibers discussed above. The chromatin exists as **heterochromatin**, which is highly condensed, and **euchromatin**, which is not so highly condensed. Human somatic cells (non-germ line) contain 22 pairs of non-identical chromosomes and 2 sex chromosomes (XX, female, and XY, male). Each eukaryotic chromosome contains one very large double-stranded DNA molecule consisting of millions of base pairs. During mitosis, all chromosomes are highly condensed and are visible in the light microscope. The ends of chromosomes are referred to as **telomeres**. The more central regions of mitotic chromosomes which become attached to spindle fibers during mitosis are called **centromeres** and contain large amounts of repetitive DNA.

Many pathological conditions result in chromosome alterations that are easily seen in mitotic chromosomes with the light microscope. These include chromosomal rearrangements, deletions, and amplifications as observed in tumors, as well as abnormal chromosome numbers and structure (trisomy of chromosome 21 in Down's syndrome and X chromosome alterations in the fragile X syndrome associated with mental retardation).

When DNA in the nucleus of an animal cell is replicated, the two strands are separated and new complementary strands are synthesized on each separated strand. The nucleosome structure must also be replicated as well. All the "old" nucleosomes remain associated with one of the separated strands and new nucleosomes (composed of newly synthesized histones from the cytoplasm) form on the other strand.

D. DNA Mutations

A gene mutation occurs when the unique sequence of deoxyribonucleotides in a gene is altered in any way. A mutation in the DNA sequence that codes for a protein can alter its amino acid sequence.

The consequences of DNA mutations on cellular metabolism can range from extremely harmful to beneficial. Mutations can occur spontaneously due enzymological errors in DNA replication, recombination, and cell division. DNA mutations can also result from environmental factors like ionizing radiation, ultraviolet light from the sun and other sources, chemical mutagens, and viruses and viral genes.

Mutagen	Type
5-bromouracil	causes transitions, replaces thymine in DNA and, after isomerization, pairs with G instead of A
aminopurine	causes transitions
hydroxylamine	causes transitions, alters C so it pairs with A
nitrous acid	causes transitions, deaminates cytosine to uracil so that it pairs with A rather than G
sulfur mustards nitrogen mustards ethyl ethanesulfonate methyl methanesulfonate	cause transitions and transversions by alkylating N7 of G, remove purines (apurination)
low pH	causes transitions, tranversions and apurination
intercalating dyes (bind between base pairs): proflavin, 5-aminoacridine, acridine orange, ethidium bromide, UV light	cause insertions and deletions resulting in frameshift mutations
ultraviolet (UV) light	dimerizes adjacent pyrimidines (thymine dimers) leading to deletions
other ionizing radiation (X-rays, radioactivity, cosmic rays)	cause severe damage to DNA by breaking covalent bonds

Table 12 - 1. Mode of Action of Mutagenic Agents

Mutations can be of a number of different kinds. **Point mutations** involve the alteration of a single nucleotide or nucleotide pair which can be of several types. A **transition** occurs when a purine is substituted for another purine or a pyrimidine is substituted for another pyrimidine. A **transversion** occurs when a purine is substituted for a pyrimidine or vice versa. A point **deletion** is the loss of a nucleotide or nucleotide pair and a point **insertion** is the gain of a nucleotide or nucleotide pair. **Gross mutations** involve the deletion or insertion of more than one nucleotide pair, up to very large segments of the genome. A **trans-**

location is the movement of a chromosomal segment to a non-homologous chromosome. **Inversions** are inverted pieces of DNA within one chromosome. Mutations which lead to defective proteins and metabolic alterations are usually deleterious to the cell. Some mutations can be harmless, leading to an amino acid change in a protein which has no effect on structure and function. Mutations can be beneficial if the mutated gene product confers a selective advantage on the organism. Table 12 - 1 is a partial list of environmental mutagens and the types of mutations they cause.

The mutation rate in organisms caused by ultraviolet and ionizing radiation should be much higher than is actually observed. This relatively low mutation rate is due in large part to the ability of organisms to repair their own DNA. **Thymine dimers** (T-T dimers), which are induced by UV irradiation, can be repaired by photoreactivation, involving the light-activation of specific enzymes (occurs in bacteria). T-T dimers are also removed by excision repair, in which an endonuclease (a nuclease which cleaves the phosphodiester bond between adjacent nucleotides) clips the phosphodiester backbone near the dimer, an exonuclease (a nuclease which cleaves phosphodiester bonds successively from an end) removes the dimer and surrounding nucleotides, and the gap is filled in by the action of DNA polymerase I and DNA ligase (Figure 12 - 4). T-T dimers can also be repaired by recombination events. In *xeroderma pigmentosum*, a human disease, there is a defect in the excision-repair enzymes causing these individuals to be susceptible to skin cancers.

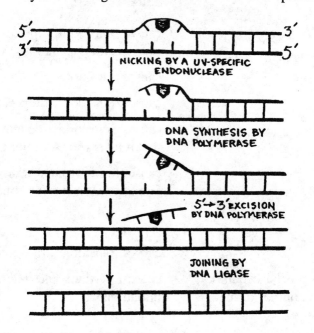

Figure 12 - 4. DNA Repair: Excision of Thymine Dimers

II. RNA

A. Structure

Ribonucleic acid (RNA) contains ribose in place of deoxyribose as the nucleotide sugar. The pyrimidine is uracil instead of thymine. RNA is usually single-stranded and contains rarely more than a few thousand nucleotides joined in phosphodiester linkages. Compared to DNA, cellular RNA is much more varied in size, structure, and function. The various kinds of RNA in the animal cell include **heterogeneous nuclear RNA (hnRNA)**, **messenger RNA (mRNA)**, **small nuclear RNA (snRNA)**, **ribosomal RNA (rRNA)**, **transfer RNA (tRNA)** and other minor RNA species. The majority of the RNA in cells is rRNA (80%), followed by tRNA (15%), mRNA (2%), and other (3%). RNA is readily hydrolyzed in alkaline solutions because of the adjacent $2'$, $3'$ OH groups on the ribose, whereas DNA (containing only the $3'$ OH group) is denatured but not hydrolyzed by alkali. RNA can contain unusual nucleotide bases like pseudouracil, ribothymidine, methyl adenine, and dihydrouracil. DNA modification in animal cells is usually limited to methylation of cytosine (5 position on base) when $5'$ to a guanosine (C^5 methylpG).

RNA is synthesized in the cell nucleus and also in mitochondria. hnRNA, which is unstable, is found in the nucleus and can be the largest of the RNA species (10 - 20,000 bases). It is the precursor of mRNA and is processed rapidly into mRNA in the nucleus. Messenger RNA (mRNA) contains the genetic information of DNA (unique sequence) in the form of the genetic code and is found primarily in the cytoplasm.

B. RNA Synthesis

1. Bacterial transcription

RNA synthesis is the process in which a single-stranded RNA is synthesized from ribonucleotide triphosphate precursors by RNA polymerase utilizing a single-stranded DNA template. The template strand of the duplex DNA molecule (bottom strand, $3'$ - $5'$ orientation) is copied into a $5'$ - $3'$ RNA transcript (transcription as well as DNA polymerization is always in the $5'$ - $3'$ direction). *In bacteria there is one RNA polymerase which transcribes all the different genes in the bacteria.* The complete enzyme (called the holoenzyme) contains 5 protein subunits: beta, beta$'$, 2 alphas, and one sigma. RNA transcripts in bacteria are not as extensively modified and processed as are transcripts in eukaryotic cells (capping, polyadenylation, and splicing.

Messenger RNAs in bacteria can be polycistronic, i.e. coding for more than one protein. In many cases, transcription of the bacterial polycistronic mRNA is under control of a promoter and operator region at which coordinated regulatory events occur. This situation is referred to as an operon. The mRNAs in eukaryotic or higher organisms tend to be monocistronic, i.e. code for only one protein. In both bacterial and higher systems, there exist accessory proteins that interact with promoter elements and RNA polymerase and help (or hinder in the case of repressor proteins) the enzyme initiate transcription. These proteins are called transcription factors.

In bacteria, transcription factors include catabolite activator protein (CAP) and the lactose operon repressor. Although the *E. coli* genome contains only one site of initiation of DNA synthesis, there exist hundreds of RNA synthesis initiation sites, each containing specific DNA sequences called **promoters** which direct RNA polymerase to the proper initiation site. In bacteria, this promoter is characterized by a AT-rich region around nucleotide position –10 from the start of transcription (at +1) and also important sequences at nucleotide position –35. Unlike DNA polymerase, RNA polymerase does not require a primer to initiate transcription. The first nucleotide transcribed is a NTP, a purine. Transcription initiation in bacteria is inhibited by the antibiotic **rifampicin**.

Transcription factors can regulate the initiation of RNA synthesis in bacterial cells. For example, the lactose repressor (which is encoded in the lactose operon) is a protein which binds to a DNA sequence called an "operator" which lies between the transcription promoter and the transcription initiation site for the beta-galactosidase gene (and two other downstream genes) which produces an enzyme which degrades lactose. If lactose is not present in the cell, then the repressor binds to the operator and blocks RNA polymerase from initiating transcription. If lactose is present in the cell, then the sugar binds to repressor, alters its conformation so that it no longer binds to the operator. RNA polymerase is now able to transcribe the galactosidase gene. This results in the production of beta galactosidase that degrades lactose, allowing the cell to use the breakdown products for metabolism. Regulation of this type is an example of **negative control**.

However, if glucose is also present, the cell prefers to metabolize glucose and the lactose operon remains turned off even in the presence of lactose. This is accomplished by a phenomenon termed **catabolite repression**. A catabolite of glucose metabolism reduces the amount of cyclic AMP (cAMP) present in the cell. Cyclic AMP is referred to as a second messenger. It is synthesized by an enzyme called adenyl cyclase which condenses the 5′ phosphate group with the 3′OH group of AMP. cAMP is degraded by a phosphodiesterase. In the absence of glucose metabolism, cAMP levels are high and the nucleotide binds and activates the catabolite activator protein, CAP. CAP when bound to cAMP is a positive transcription factor for the lactose operon. That is, it binds upstream of the lactose promoter and facilitates binding and transcription by RNA polymerase when lactose is present (so the lactose repressor is not blocking RNA polymerase binding). Therefore if only lactose is present, the lactose operon is turned on transcriptionally but if glucose is also present, the operon will remain turned off because CAP will not activate transcription under these conditions. This is an example of **positive control**.

Transcription termination in bacteria proceeds primarily via two mechanisms. In some genes, the newly synthesized RNA at the end of a gene undergoes intrastrand base-pairing forming a hairpin structure. This hairpin is usually followed by a string of U's. This structure triggers release of the RNA polymerase from the DNA template and release of the RNA chain. Other bacterial genes require a protein factor termed **rho** for efficient transcription termination.

In some amino acid biosynthetic operons a form of transcription termination occurs under certain conditions and is referred to as **attenuation**. In the histidine biosynthetic operon, there exists a leader RNA sequence positioned before the protein coding regions of the operon that contains 7 contiguous codons for histidine. In the absence of histidine, the operon is transcribed by RNA polymerase and ribosomes bind the nascent RNA chain and begin translation in a process called coupled transcription-translation. Because histidine is absent (and also charged histidyl-tRNA), the ribosome stalls at the 7 histidine codons and the leader RNA assumes a folded structure that allows RNA polymerase to continue transcription of the operon. However, in the presence of excess histidine, the ribosome translates through the leader mRNA, altering its structure so as to produce a hairpin recognized as a termination signal by RNA polymerase moving just ahead of the ribosome. The result is the transcription is terminated under conditions where new histidine synthesis is not needed.

2. Mammalian transcription

RNA synthesis in animal cells appears to be more complicated than in bacteria, possibly because there are three distinct RNA polymerases, RNA pol I, II, III, a multitude of transcription factors, and associated RNA processing events.

RNA pol I transcribes the 18S and 28S rRNAs in an organelle in the nucleus called the **nucleolus** (a 45S precursor rRNA is transcribed in the nucleolus and then processed to form the 18S and 28S and 5.8S rRNAs). **RNA pol II** transcribes mRNA-coding genes. RNA pol II also transcribes the majority of the snRNAs. **RNA pol III** transcribes the 5S rRNA genes and tRNA genes and also the 7S RNA found in the signal recognition particle. The animal cell RNA polymerases exhibit different sensitivities to the mushroom-derived toxin, **alpha amanitin**. RNA pol I is resistant to this toxin; RNA pol II is very sensitive and RNA pol III is moderately sensitive.

The DNA promoter sequences for these enzymes are distinct. Pol I genes (coding for rRNA) contain a promoter near the +1 start site whereas genes transcribed by Pol III (coding for tRNA) contain promoters within the genes themselves. Most pol II (mRNA) transcription promoters include a "TATA" box at about −35 position. In addition, upstream activating sequences (CAAT box, GC box) can be located from −70 to −100. Also many pol II genes contain **enhancer** sequences which can be located thousands of base pairs away from the +1 transcription start site. The addition of the cap structure to the +1 nucleotide occurs very soon after the initiation of transcription. Poly A tails are added to the 3' end of the transcript after a cleavage has occurred in the RNA transcript at an AU rich region. Pol I and III initiate with a pppG and pol III is capable of termination by itself. Pol II transcribes a hnRNA molecule which is then processed into mRNA by a process called splicing.

Although bacterial transcription factors have the helix-turn-helix DNA binding motif (a protein domain involved in DNA recognition), eukaryotic transcription factors have other motifs in addition to the helix-turn-helix. These include the "zinc finger" and "leucine zipper" motifs.

III. REVIEW QUESTIONS ON NUCLEIC ACIDS

DIRECTIONS: For each of the following multiple-choice questions (1 - 42), choose the ONE BEST answer.

1. DNA ligase joins segments of DNA:

A. forming a β-glycosidic linkage between the preceding base and the next sugar
B. catalyzing the formation of a phosphodiester bond between a free 5'-nucleotide phosphate group and a 2'-sugar hydroxyl group
C. inserting an RNA primer between adjacent segments
D. catalyzing the formation of a phosphodiester bond between a free 3'-hydroxyl at the end of one DNA chain and a 5'-phosphate at the end of the other DNA chain
E. sealing the gaps between RNA primers and the growing DNA strands

2. Nucleosomes are:

A. found primarily in prokaryotic cells and not in mammalian cells
B. nucleoprotein particles of DNA wrapped around a core of basic proteins containing a high proportion of lysine and arginine
C. nucleoprotein particles of RNA wrapped around a core of eight histone proteins
D. particles in which the histones are bound to the nucleic acid by a phosphodiester bond
E. nucleoprotein particles in which the amino acids are negatively charged at neutral pH

3. It was established that DNA is replicated semi-conservatively by demonstrating that:

A. only part of the DNA is replicated at any particular time
B. during replication none of the genetic information is conserved
C. after replication each daughter molecule contains one parental strand and one new strand
D. after replication both parental strands were in one daughter cell and both new strands were in the other daughter cell
E. the mechanism of DNA replication involved 5' → 3' synthesis of one parental strand and 3' → 5' synthesis of the other strand.

4. An example of a palindrome in a DNA sequence is:

A. ATGCCG
   TACGGC

B. GGCCGG
   CCGGCC

C. CTAGGG
   GATCCC

D. GAATTC
   CTTAAG

E. TCTGAC
   AGACTG

5. The sequence of a short duplex DNA is

5′ GAACCTAC 3′
3′ CT TGGATG 5′.

What is the corresponding mRNA sequence transcribed from this region?

A. GAACCTAC
B. CUUGGAUG
C. CAUCCAAG
D. GAACCUAC
E. GUAGGUUC

6. RNA polymerases require as substrates:

A. ATP, GTP, TTP, CTP
B. ATP, GTP, UTP, CTP
C. dATP, dGTP, dTTP, dCTP
D. an RNA template
E. ribosomes

7. By which of the following means are bases linked to the pentoses in RNA and DNA molecules?

A. They are linked together by β-glycosidic bonds.
B. They are linked together by 3′, 5′ phosphodiester bridges.
C. They are linked together by 2′, 5′ phosphodiester bridges.
D. They are linked together by α-glycosidic bonds.
E. They are held together by electrostatic attraction.

8. DNA polymerase I is extremely accurate in its ability to make a complementary copy of the template. The key activity that allows the polymerase to be so accurate is the:

A. ligase activity
B. adenosine deaminase activity
C. gyrase activity
D. 3′ → 5′ exonuclease activity
E. 5′ → 3′ exonuclease activity

9. Repair of single-stranded nicks in double stranded DNA and formation of circular molecules from linear DNA pieces can be accomplished by:

A. helicases
B. nucleases
C. restriction enzymes
D. ligases
E. polymerases

10. Restriction enzymes:

A. cut single-stranded DNA in a random fashion
B. cut double-stranded DNA in a sequence-specific fashion
C. are used to join two DNA molecules together
D. are not affected by DNA methylation
E. require ATP as a cofactor

11. A point mutation in a mRNA codon will most likely cause:

A. mRNA degradation
B. inactivation of ribosomes
C. an altered amino acid sequence
D. incomplete transcription
E. inhibition of splicing

12. The nucleosome core contains:

A. 1 copy each of histone H2A, H2B, H3, and H4.
B. 2 copies each of histone H2A, H2B, H3, and H4.
C. DNA covalently linked to histones
D. histone H1
E. RNA polymerase

13. When two DNA molecules are compared according to buoyant density and their "melting points" (T_m), the one with the greatest A - T content will have:

A. the higher density and higher T_m
B. the lower density and lower T_m
C. the higher density and lower T_m
D. the lower density and higher T_m
E. no detectable difference

14. In eukaryotic (mammalian-like) cells DNA molecules exist as nucleoproteins. The protein components are primarily:

A. histones
B. glycoproteins
C. lysine
D. arginine
E. albumin

15. In polynucleotides the individual nucleosides are linked together with:

A. hydrogen bonds
B. ionic bonds
C. glycosidic bonds
D. phosphodiester bonds
E. phosphomonoester bonds

16. Which nucleic acid is most rapidly degraded in mammalian cells?

A. rRNA
B. hnRNA
C. tRNA
D. rDNA
E. 5S RNA

17. Which nucleic acid is most stable in mammalian cells?

A. mRNA
B. 45S rRNA
C. duplex DNA
D. thymine dimers
E. precursor tRNA

18. Denatured human lymphocyte DNA will not hybridize with human:

A. lymphocyte rRNA
B. kidney tRNA
C. denatured mitochondrial DNA
D. denatured liver DNA
E. brain mRNA.

19. Tissue cells incubated with [^{3}H] thymidine will most rapidly accumulate the radioactive isotope in:

A. mitochondria
B. ribosomes
C. hnRNA
D. rough ER
E. nuclei

20. DNA polymerase utilizes which compounds as substrates?

A. CTP, UTP, GTP, ATP
B. dCTP, dUTP, dGTP, dATP
C. dATP, dCTP, dTTP, dGTP
D. deoxyribonucleotide diphosphates
E. ribonucleotide diphosphates

21. In the Watson-Crick model for DNA, the molecule is a(an):

A. single stranded helix
B. α-helical structure
C. right handed, double-stranded helix of the same polarity
D. left-handed, double-stranded helix of opposite polarity
E. right-handed, double-stranded helix of opposite polarity

22. A codon mutation resulting from a deletion of a nucleotide base results in a:

A. translocation
B. inversion
C. frameshift change
D. transversion
E. transition

23. At the growing fork of replicating DNA small DNA fragments are found and are referred to as:

A. Watson fragments
B. Cairn fragments
C. Meselson and Stahl fragments
D. Okazaki fragments
E. Avery fragments

24. DNA and RNA molecules in solution absorb ultraviolet light strongly at 260 nm. This absorption is due to:

A. pyrimidines
B. purines
C. both pyrimidines and purines
D. phosphate
E. ribose and deoxyribose

25. When the temperature of a DNA solution is raised high enough the two strands will unwind from each other. This is called:

A. renaturation
B. hybridization
C. hydrolysis
D. denaturation
E. condensation

26. Which of the following is NEITHER a substrate NOR a cofactor for DNA polymerase III?

A. Mg^{++}
B. dCTP
C. UTP
D. template DNA
E. RNA primer

27. A promoter:

A. contains conserved sequences upstream from the beginning of a gene to which RNA polymerase binds.
B. is a binding site for DNA polymerase I
C. is a nick in a DNA duplex
D. is the site at which DNA replication begins
E. is required by DNA polymerase III

28. Replication is a sequential process. In the following list of the proteins involved in replication, which one is NOT in proper sequence?

A. helicase
B. RNA polymerase
C. DNA ligase
D. DNA polymerase III
E. DNA polymerase I

29. The AIDS virus (HIV) contains an RNA genome. Which enzyme is responsible for converting this genome into DNA in T lymphocytes?

A. Ligase
B. EcoRI
C. DNA polymerase
D. Reverse transcriptase
E. RNA polymerase

30. RNA polymerase is a multi-component enzyme made up of several polypeptide subunits. The subunit that recognizes and binds to the promoter sequence on the DNA template is referred to as:

A. omega factor
B. rho factor
C. alpha subunit
D. beta subunit
E. sigma factor

31. EcoRI cleaves:

A. single-stranded DNA
B. a DNA palindrome
C. double-stranded RNA
D. purine base from deoxyribose
E. at all four base sequences

32. New DNA strands use RNA sequences as primers. These are then extended by:

A. DNA polymerase I
B. DNA ligase
C. DNA polymerase III
D. RNA polymerase
E. Helicase

- 197 -

33. Mammalian RNA polymerase I synthesizes:

A. 5S RNA
B. tRNA
C. mRNA
D. both tRNA and 5S RNA
E. rRNA

34. The top and bottom strands of DNA are:

A. not complementary
B. parallel
C. anti-parallel
D. held together by peptide bonds
E. held together by ionic bonds

35. RNA is transcribed in the (referring to the RNA):

A. 5' to 3' direction
B. 3' to 5' direction
C. amino-terminal to carboxy-terminal direction.
D. carboxy-terminal to amino-terminal direction.

36. Genetic information in retroviruses flows from:

A. DNA to RNA to protein
B. RNA to DNA to RNA to protein
C. DNA to RNA to DNA to protein
D. RNA to DNA to protein.
E. protein to RNA to DNA

37. Base pairing rules state that in DNA:

A. A = U
B. G = T
C. A = C
D. A = T
E. C = U

38. The 3' end of a nucleic acid strand contains:

A. an amino group
B. a phosphate group
C. a β-lactam group
D. a hydroxyl group
E. a triphosphate group

39. DNA contains:

A. pyrimidine ribose monophosphates
B. purine deoxyribose diphosphates
C. purine deoxyribose monophosphates
D. purine deoxyhexose monophosphates
E. 5'-2' phosphodiester bonds

40. The DNA consensus sequence "TATA" is a:

A. initiation signal for DNA synthesis
B. initiation signal for RNA synthesis
C. histone binding site
D. ribosome binding site
E. DNA methylation site

41. Histones are rich in the amino acid(s):

A. cysteine
B. phenylalanine
C. tryptophan
D. leucine and valine
E. lysine and arginine

42. The DNA template is read _____; and the new strand grows _____.

A. 3'-5' ... 3'-5'
B. 5'-3' ... 5'-3'
C. 5'-3' ... 3'-5'
D. 3'-5' ... 5'-3'

IV. ANSWERS TO QUESTIONS ON NUCLEIC ACIDS

1. D	15. D	29. D
2. B	16. B	30. E
3. C	17. C	31. B
4. D	18. C	32. C
5. D	19. E	33. E
6. B	20. C	34. C
7. A	21. E	35. A
8. D	22. C	36. B
9. D	23. D	37. D
10. B	24. C	38. D
11. C	25. D	39. C
12. B	26. C	40. B
13. B	27. A	41. E
14. A	28. C	42. D

13. PROTEIN BIOSYNTHESIS

Jay Hanas and Albert M. Chandler

I. INTRODUCTION

A protein consists of 20 different amino acids linked head to tail by peptide bonds. The amino acids are arranged in a specific sequence in each particular protein. In order to accomplish this feat each cell must have: (1) cellular machinery by which the proteins are assembled; (2) informational mechanisms to designate the amino acid sequences; (3) enzymes and other factors to carry out the detailed assembly processes.

II. CELLULAR MACHINERY

Protein biosynthesis involves three major species of RNA: **ribosomal RNA (rRNA)**, **messenger RNA (mRNA)** and **transfer RNA (tRNA)**. In addition, in eukaryotic organisms, two other types, **heterogeneous nuclear RNA (hnRNA)** and **small nuclear RNA (snRNA)**, are also involved.

A. Ribosomal RNA, Ribosomes and Polysomes

If the post-mitochondrial supernatant from a mammalian cell source is centrifuged at 105,000 x g for 1 hour, a pellet is formed which is termed the **microsomal** fraction. This consists of the membranes of the **endoplasmic reticulum**. Membranes to which ribosomes are attached by their larger subunits are called **rough endoplasmic reticulum (RER)**. Membranes lacking these particles are termed **smooth endoplasmic reticulum (SER)**. In addition, free ribosomes will be found both as individual particles and attached together in clusters by a thread of messenger RNA. These clusters are called **polyribosomes** or **polysomes**. *Protein synthesis always takes place on polysomes.* As a general rule, proteins destined to be exported from the cell or destined to become part of cell membranes are synthesized on membrane-bound polysomes. Proteins destined to stay within the cell, i.e., "housekeeping proteins", are synthesized on free polysomes. The ribosomes and polysomes of bacteria are usually found free, seldom membrane-bound.

1. Structure of ribosomes

Bacterial. Individual ribosomes, when examined with an electron microscope, will be seen as spherical particles about 180 - 220 Å in diameter and consisting of two subunits. An example of the structure and composition of *E. coli* ribosomes is shown in Figure 13 - 1. The whole ribosome sediments with a value of 70S and can be broken down into a 50S subunit and a 30S subunit. "S" refers to the sedimentation constant and is indicative of the size of the RNA. Transfer RNA, for example, sediments during centrifugation as a 4S entity. The 50S subunit yields a characteristic 23S and a 5S rRNA while the 30S subunit yields a 16S rRNA. Both subunits yield a characteristic set of specific ribosomal proteins, 21 from the 30S and 31 from the 50S.

Mammalian. The ribosomal RNAs in eukaryotic organisms have a sedimentation constant of 80S made up of 60S and 40S subunits. The 60S subunit contains three major rRNA species, the 5S, 5.8S, and 28S RNAs and 49 proteins, while the 40S subunit contains one 18S rRNA and 33 proteins. The 5S rRNA contains 120 nucleotides, the 5.8S rRNA contains 160 nucleotides, the 18S rRNA contains about 1900 nucleotides and the 28S rRNA contains about 5000 nucleotides (Figure 13 - 1).

2. Synthesis (Figure 13 - 3)

Synthesis of eukaryotic ribosomes occurs in the nucleus and involves a special organelle, the **nucleolus**. Synthesis begins with production in the nucleolus of a large **45S precursor** by the enzyme **RNA polymerase I.**

PROKARYOTIC RIBOSOME (*E. coli*)   EUKARYOTIC RIBOSOME (cytoplasmic)

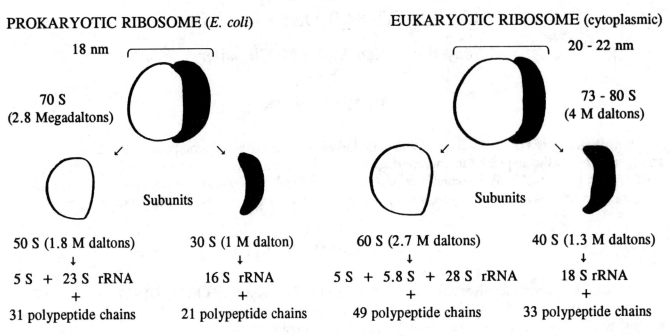

Figure 13 - 1. Composition of Prokaryotic and Eukaryotic Ribosomes.

After a series of alkylations and cleavages this single precursor gives rise to the 28S, 18S, and 5.8S species of rRNA. (The clear areas in Figure 13 - 3 are portions of the original 45S molecule that are degraded). 5S rRNA is made elsewhere in the nucleus. Ribosomal RNA also contains methylated bases and sugars.

Mitochondria and chloroplasts of eukaryotic cells contain their own genetic systems and are capable of autonomous protein synthesis. They also have ribosomes that appear to be intermediate between cytoplasmic ribosomes of eukaryotes and those of bacteria but differ from both in size and RNA composition. Mitochondrial ribosomes are smaller than cytoplasmic ribosomes, having a sedimentation constant of 55S. *In biological behavior they closely resemble bacterial ribosomes.*

B. Messenger RNA

The diagram shown below (Figure 13 - 2) is a schematic representation of the general linear structure of a typical eukaryotic mRNA molecule.

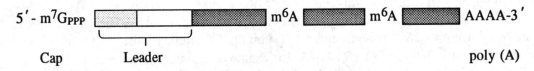

Figure 13 - 2. Structure of a Typical Eukaryotic Messenger RNA.

A "**cap**" is essential for the initiation of translation and acts to protect the molecule from degradation. The cap consists of a GTP molecule methylated in the 7 position of the guanine moiety and attached through the third phosphate residue $5'-5'$ to the end of the mRNA. If the cap is removed or if it is not methylated the mRNA will either not be translated at all or will be translated with markedly reduced efficiency. The "leader" is recognized by binding sites on the ribosome and by certain initiation factors. The function of the internal methylations is unknown. The **poly(A) "tail"** confers stability on the molecule and is suspected to be involved in the transport of newly synthesized mRNA from the nucleus to the cytoplasm. The poly(A) tail can range from 100 - 200 nucleotides long. There is no poly(T) region on the DNA template coding for the tail. The tail is synthesized from ATP molecules which are added sequentially by *poly(A) polymerase*, an

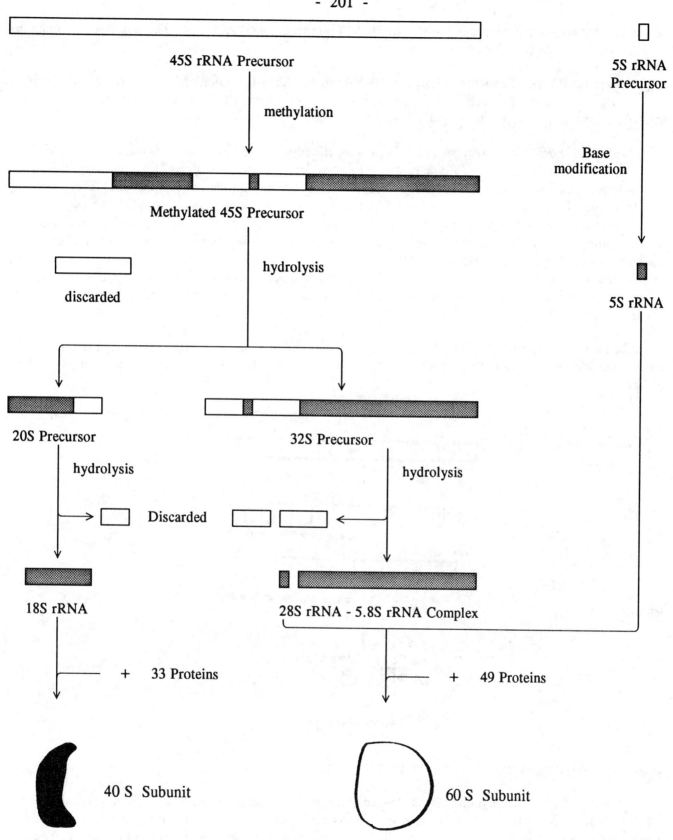

Figure 13 - 3. Assembly of Eukaryotic Ribosomes.

enzyme that recognizes a unique sequence (AAUAAA) on the primary transcript. The polymerase can be inhibited by **cordycepin** (3′ - deoxyribose).

Bacterial mRNAs differ from eukaryotic mRNAs in that they have neither cap structures nor poly(A) tails. As a consequence bacterial mRNAs have very short half lives, (2 - 3 minutes), whereas mammalian mRNAs can have comparatively long half lives (hours or days).

Synthesis of mRNA in Eukaryotic cells. In prokaryotic cells the coding regions of the DNA usually form one long continuous sequence. In eukaryotic cells, however, the coding regions are usually "split", that is, coding sequences (**exons**) are separated from one another by non-coding sequences (**introns**) (Figure 13 - 4). The functions of these introns are unknown. When the gene is transcribed, the primary RNA transcript contains sequences from both exons and introns. The introns must be removed and the exons "spliced" together to form functional mRNA. This splicing process is carried out by ribonucleoprotein particles called **small nuclear ribonucleoprotein particles (snRNPs)** which come together to form complex structures called **spliceosomes**. The RNA components of these particles are small, stable RNAs designated U1, U2, . . . U8, ranging in size from 150 - 300 bases. Several of these RNAs appear to be involved in the processes converting hnRNA to mRNA.

The introns are removed as "lariat" structures, in which the 5′ end of the intron is covalently attached to an internal adenosine residue to form a loop with a tail. All introns have **GU** at their 5′ ends and **AG** at their 3′ ends.

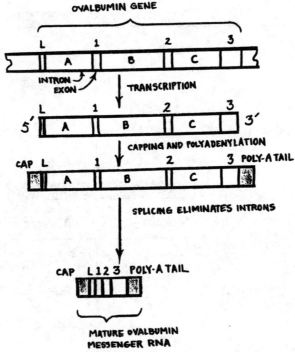

Figure 13 - 4. Structure of the Ovalbumin Gene.

Patients with the autoimmune disease **systemic lupus erythematosus (SLE)** and other autoimmune diseases produce autoantibodies against these snRNPs as well as against other cellular nucleoproteins. Certain mutations in human globin genes will not allow the proper splicing of primary transcripts and can be the cause of **thalassemias**.

C. Transfer RNA

There are 50 - 60 different, but related, species. They are all similar in structure, but do have differences in base composition and structure, in amino acid specificity and in codon recognition ability. They contain 10 - 20% unusual "minor" bases. These minor bases are formed after the polynucleotide chain has been transcribed. Each reaction requires a specific enzyme.

Transfer RNAs are 70 - 90 nucleotides in length; M.W. about 25 - 30,000 and all can be folded into a "cloverleaf" secondary structure. Three-dimensionally they are "L" shaped. The structure of tRNA is very well understood due to X-ray crystallography as illustrated below:

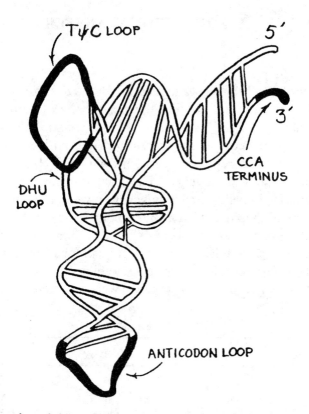

Figure 13 - 5. Three-dimensional Structure of a Representative tRNA.

All end in -**CCA** at the 3′ end. The amino acid is bound to the ribose unit of the terminal adenosine. The "amino acid arm" contains 7 base pairs, the "anticodon arm" contains 5 base pairs and the anticodon loop always contains 7 nucleotides. Thus, all tRNAs are of the same length. All contain a TUC loop and a dihydrouracil loop which are close together in the 3-D structure of the molecule but far apart in the 2-D structure. These loops contain modified bases. These probably associate with specific binding sites on the ribosome. The **anticodon** (codon recognition site) usually has on one side a pyrimidine (U) and on the other an alkylated purine. The codon and anticodon always pair up in an antiparallel manner.

D. Amino Acid Activation

In order to be coupled into a growing polypeptide chain an amino acid must first be "activated", that is, coupled to a specific tRNA. This coupling requires specific *amino acid activating enzymes* also known as *aminoacyl-tRNA synthetases*. There is at least one amino acid activating enzyme for each amino acid. These enzymes have double specificity; that is, they have two specific binding sites, one for a specific amino acid and the other for a specific tRNA. They catalyze the following reactions (next page). Although these are reversible *in vitro*, because of the presence of very active inorganic pyrophosphatases in living cells, the overall reaction favors the formation of aminoacyl-tRNAs. The amino acid is coupled via its carboxyl group in ester linkage with the **3′OH** of the ribose moiety of the terminal adenosine moiety of the tRNA.

(1) $Enz_1 + ATP + AA_1 \longrightarrow Enz_1\text{-}AA_1\text{-}AMP + PP_i$

(2) $Enz_1\text{-}AA_1\text{-}AMP + tRNA_1 \longrightarrow AA_1\text{-}tRNA_1 + Enz_1 + AMP$

Sum: $AA_1 + tRNA_1 + ATP \longrightarrow AA_1\text{-}tRNA_1 + AMP + PP_i$

III. INFORMATION TRANSFER

A. The Genetic Code

The genetic code is the relationship between the sequence of bases in DNA (or its RNA transcripts) and the sequence of amino acids in a protein. Deciphering the genetic code in the 1960's was a monumental achievement, ushering in the modern era of biology. Although genetic information is stored in DNA in unique sequences of the four deoxynucleotides and transcribed into an identical RNA sequence of four ribonucleotides (deoxythymidylate in DNA replaced by ribouridylate in RNA), this information must then be translated into a unique protein amino acid sequence. There are 20 distinct amino acids commonly found in proteins, linear arrays of amino acids joined by amide linkages between the alpha amino group of one amino acid and the alpha carboxyl group of the next. Given four unique bases (A, U, G, or C), a singlet or doublet genetic code would only give 4 or 8 combinations (4^1 or 4^2), not enough to code for 20 amino acids. A triplet code (4^3) would code for 64 unique possibilities, more than enough for 20 amino acids. By using synthetic ribonucleotide polymers in crude *in vitro* protein synthesizing systems the genetic code was indeed found to be **triplet** in nature and **codons** (three adjacent ribonucleotides) were assigned to all 20 amino acids.

1st pos. 5′ end	2nd position				3rd pos. 3′ end
	U	C	A	G	
U	Phe	Ser	Tyr	Cys	U
	Phe	Ser	Tyr	Cys	C
	Leu	Ser	**Stop**	**Stop**	A
	Leu	Ser	**Stop**	Trp	G
C	Leu	Pro	His	Arg	U
	Leu	Pro	His	Arg	C
	Leu	Pro	Gln	Arg	A
	Leu	Pro	Gln	Arg	G
A	Ile	Thr	Asn	Ser	U
	Ile	Thr	Asn	Ser	C
	Ile	Thr	Lys	Arg	A
	Met	Thr	Lys	Arg	G
G	Val	Ala	Asp	Gly	U
	Val	Ala	Asp	Gly	C
	Val	Ala	Glu	Gly	A
	Val	Ala	Glu	Gly	G

Table 13 - 1. The Genetic Code.

B. General Features of the Code

The code is **degenerate**; that means that in many cases, several different codons code for the same amino acid. During protein synthesis the anticodons of charged tRNAs base pair with the codons in mRNA. In fact some charged tRNAs can bind to different codons due to a phenomenon called **"wobble"** in which non-Watson-Crick base pairing can occur between the third position of some codons and the first position of a tRNA anticodon. The code is **non-overlapping**; one codon never codes for more than one amino acid. The code **lacks punctuation**; mRNA is translated by ribosomes in one continuous process, three nucleotides at a time. If a mutation occurs in the mRNA such that one nucleotide has either been deleted or added, the reading phase is changed accordingly (a **"frameshift"** mutation). Once changed, the phase will not change unless another compensating mutation occurs downstream. Three triplets in the genetic code were found not to code for amino acids (viz. **UAA, UAG,** and **UGA**). These codons were later found to signal termination of protein synthesis and are referred to as **termination, stop** or **nonsense** codons. The codon, **AUG,** codes for **methionine** and is also used as the **initiation** codon.

With the exception of mitochondria, the code is universal. That is, all forms of life as well as viruses have the same codon/amino acid assignments. Mitochondria have some minor changes in the codon/amino acid assignments indicating a different evolutionary history for these organelles.

IV. PEPTIDE BOND FORMATION

A. Prokaryotes

Protein synthesis can be divided into three steps, **initiation, elongation,** and **termination.** All prokaryotic proteins begin with N-formyl methionine (fMet). The initiation step of protein synthesis places the initiator codon of the mRNA (usually AUG, sometimes GUG in prokaryotes) and the fMet-tRNA at the proper site on the ribosome for the start of translation. The 5′ leader sequence of mRNA in bacteria has a sequence which is complementary to the 3′ end of the 16S RNA in the 30S ribosomal subunit (**Shine-Delgarno sequence**). Base pairing occurs between these complementary regions and assists in the proper placement of the mRNA on the 30S subunit.

1. Chain Initiation

a. *The role of N-formylmethionine tRNA$_f$.* The methionine-accepting tRNAs consist of two distinct species, termed **tRNA$_f^{Met}$** and **tRNA$_m^{Met}$.** These can undergo the following series of reactions in prokaryotic cells:

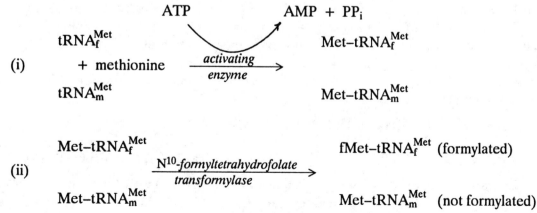

Both Met-tRNAs recognize AUG as a codon, but fMet-tRNA$_f$ is used only when AUG (or GUG) is in the chain initiation position, and Met-tRNA$_m$ only for chain elongation. The formyl group is attached to the α-amino group of the methionine residue. fMet-tRNA$_f$ is used to initiate protein synthesis in all prokaryotes (bacteria, viruses) and in mitochondria and chloroplasts of higher organisms.

All ribosomes have two aminoacyl-tRNA binding sites on their surfaces. One of the sites is called the **acceptor site (A-site)** and the other the **peptidyl site (P-site)**. These binding sites extend over both the large and small ribosomal subunits.

b. *Initiation Factors*. There are three protein factors involved in initiation. These are called **IF-1**, **IF-2** and **IF-3**.

c. *Initiation Steps* (Figure 13 - 6). Binding of mRNA to the initiation factor-30S subunit complex releases IF-3. Next, the 50S subunit binds to the 30S subunit, mRNA, fMet-tRNA, IF-1, IF-2, GTP complex. IF-1 and IF-2 are released at this stage upon GTP hydrolysis. The fMet-tRNA is positioned correctly (anticodon bound to initiator AUG in mRNA) in the P (peptidyl tRNA) site of the 70S ribosome, poised for the elongation phase of protein synthesis.

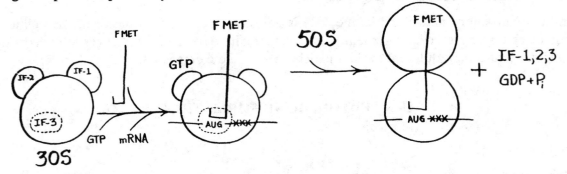

Figure 13 - 6. Initiation Phase.

2. Chain Elongation

a. *Elongation Factors*. There are three elongation factors termed **EF-Tu, EF-Ts, EF-G**. In addition, there is a catalytic activity on the 50S subunit called *peptidyl transferase* which actually forms the peptide bond. This activity is activated only when the 50S subunit is combined with the 30S subunit on the polysome.

b. *Elongation Steps* (Figure 13 - 7). Aminoacyl tRNA binding depends upon the matching of its anticodon with the triplet codon of the mRNA positioned in the A site. Charged tRNA is complexed with elongation factor Tu and GTP. Once the charged tRNA is bound to the ribosome, the GTP is hydrolyzed to GDP and Tu is released. Tu is recycled with GTP by factor Ts. Peptide bond formation between the amino group of the aminoacyl tRNA (amino group condensed with the tRNA 3´OH) in the A site and the carboxyl group of the peptidyl tRNA in the P site is catalyzed by peptidyl transferase, an enzyme located on the 50S subunit. Once peptide bond formation occurs, the tRNA in the A-site now containing the growing polypeptide chain is transferred to the P-site along with concomitant release of the deacylated tRNA in the P-site. This reaction is promoted by elongation factor G and GTP. GTP hydrolysis results in the release of G factor. The A site is now vacant and ready for another elongation round. In this manner, the polypeptide chain grows from N-terminal to C-terminal.

c. *Recycling of EF-Tu* occurs as follows:

$$[EF\text{-}Tu]\text{-}GDP + GTP + aa\text{-}tRNA \xrightarrow{EF\text{-}Ts} [aa\text{-}tRNA]\text{-}[EF\text{-}Tu]\text{-}GTP + GDP$$

d. *Correction of errors*. The error frequency of protein synthesis (incorporation of an amino acid from an aminoacyl tRNA that was "misread" by the ribosome) is extremely low. This is due to proofreading of the peptidyl-tRNA bound to the P-site in the ribosome. The ribosome checks the match of the codon-

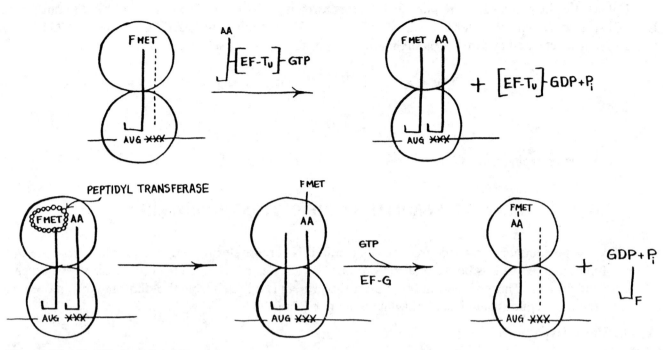

Figure 13 - 7. Elongation Phase.

anticodon interaction in the P-site and, if it is not correct, ejects the peptidyl tRNA from the ribosome in a GTP-dependent process. The antibiotic streptomycin interferes with this proofreading and causes misreading of the genetic code.

3. Chain Termination

After many elongation cycles the ribosome eventually encounters one or more chain termination codons (**UAA, UAG** or **UGA**) on the mRNA. There are three release factors, **RF-1, RF-2** and **RF-3**, involved in chain termination in prokaryotes. These factors can bind to one or more termination codons in such a manner that the specificity of the 50S-bound peptidyltransferase is altered so that the peptide chain is transferred to water instead of to another α-amino group. The protein molecule is released at this stage. GTP hydrolysis results in the release of the termination factors.

B. Eukaryotes

The process of peptide bond formation on mammalian ribosomes is very similar to that for bacteria. Differences do exist, however. Eukaryotic mRNAs do not have Shine-Delgarno sequences. Chain initiation requires Met-tRNA$_i$ and the methionine is NOT formylated. There are at least thirteen initiation factors instead of three and in addition to GTP, ATP is required. The initiation factors in eukaryotes all have the prefix eIF-. Although all the initiation factors are important, two should be pointed out. The first is **eIF-4B**, the **cap recognition factor**. This factor, with the assistance of other proteins binds the methylated cap of the mammalian mRNA to the surface of the ribosome which then "scans" the message in the 3′ direction until it encounters the first AUG which serves as the chain initiation codon.

The second is **eIF-2** which forms a **ternary** complex with GTP and Met-tRNA$_i$. This complex must be formed before it can associate with the 40S ribosomal subunit. Failure to form the ternary complex prevents chain initiation. If eIF-2 is phosphorylated by certain protein kinases, it can no longer form a viable ternary complex and protein synthesis is inhibited. This is observed in the action of **Heme Controlled Repressor (HCR)** in reticulocytes and as part of the mechanism of action of certain **interferons**.

Eukaryotic chain elongation requires two factors termed **eEF-1** (Transferase I) and **eEF-2** (Transferase II). eEF-1 is equivalent to bacterial EF-Tu and EF-Ts; eEF-2 is equivalent to EF-G. eEF-2 but NOT EF-G is irreversibly inactivated by **diphtheria toxin** according to the following scheme:

$$\text{eEF-2 + NAD} \xrightarrow{\text{toxin}} \text{eEF-2 + Niacin}$$
$$|$$
$$\text{ADP}$$
$$\text{(inactive)}$$

In eukaryotes chain termination requires only one release factor, **RF**.

V. POSTRIBOSOMAL MODIFICATION OF PROTEINS

Very few proteins when isolated from cells have their N-termini starting with methionine and none with N-formyl-methionine. This is because of the presence of a series of specific aminopeptidases which cleave off these end groups. This process can occur after the protein is released from the ribosome or can even occur while the polypeptide chain is still growing on the ribosome.

A. The Signal Hypothesis

Proteins destined for export from the cell or destined to become membrane components usually have a specific N-terminal sequence, hydrophobic in character, called a "signal" sequence which usually is located at the amino terminus of the protein (about the first 20 amino acids). This acts as a "signal" for the insertion of the growing polypeptide chain through the lipid bilayer of the endoplasmic reticulum membrane. According to the "signal hypothesis", when a translating ribosome on a polysome has synthesized enough of the protein such that its 20 amino acid signal sequence emerges from the large subunit, the signal sequence is recognized by the **signal recognition particle (SRP)** The scheme shown in Figure 13 - 8 is a highly simplified version of **Blobel's signal hypothesis**.

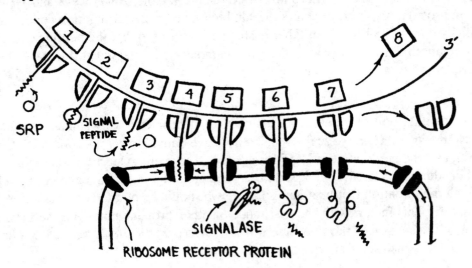

Figure 13 - 8. Secretion of Proteins through the Membranes of the RER.

The SRP is a ribonucleoprotein containing a 7S RNA and 6 - 8 specific proteins. The SRP binds to the signal sequence and also to the translating ribosome which results in a halt of protein synthesis. The SRP-ribosome complex is then recognized by a **docking protein** located on the surface of the rough endoplasmic reticulum. This docking protein binds the SRP-ribosome complex to the membrane resulting in the release of SRP and the resumption of translation. Translation causes the amino terminal signal sequence to be inserted through the lipid bilayer into the lumen of the rough ER. The protein is transported inside the rough ER by the translation process. During this process, the signal sequence is usually cleaved off.

B. Glycosylation of Proteins

The large majority of proteins synthesized in eukaryotic cells undergo glycosylation after the polypeptide backbone has been synthesized. The oligosaccharide side chains of these glycoproteins are formed in two stages. In the RER, a "mannose core", consisting of two N-acetylglucosamine, nine mannose and three glucose residues, is attached to the nascent glycoprotein after the polypeptide chain has been inserted into the lumen of the RER. The mannose core is assembled in the cytosol on a isoprenoid lipid called **dolichol phosphate**. The core is transferred from the dolichol phosphate, through the membrane, to the amide group of an asparagine residue. This process can be inhibited in animal cells by **tunicamycin** which blocks the formation of dolichol-PP-oligosaccharides.

While in the lumen of the RER the mannose core begins to be processed, i.e., certain glucose and mannose residues are removed. The glycopeptide is then encased in vesicles pinched off from the RER and transported to the *cis*-side of the **Golgi apparatus** where further processing occurs within the Golgi membranes. As the glycopeptide traverses the *cis*, medial and *trans* structures of the Golgi apparatus, extensive remodeling of the oligosaccharide side chains occur and further peripheral sugars are attached until the glycoprotein is in finished form.

The completed glycoproteins are then enclosed in vesicles which are transported from the *trans* face of the Golgi apparatus with the aid of microtubules to the inner surface of the plasma membrane. The vesicles fuse with the plasma membrane and release their contents into the intercellular fluid.

Glycoproteins destined to become **membrane** components have regions consisting of hydrophobic amino acids which are inserted into the lipid bilayer of the plasma membrane anchoring the glycoproteins in the membrane. Glycoproteins destined to become **lysosomal** enzymes have certain mannose residues phosphorylated in the 6 position in the membranes of the *cis* Golgi. This is a signal that these proteins are to be transported to pre-lysosomal vesicles. The phosphorylation is a two step process; a transferase first attaches N-acetyl glucosaminyl phosphate to the mannose and then the N-acetyl glucosamine is removed, leaving the phosphate attached to the mannose residue. A genetic deficiency in the transferase is the cause of **I-disease**.

The role of the oligosaccharide side chains appears to be topological, that is, they offer further structural components that can act as signals for binding to receptors or to other proteins. They may also control protein folding and the life-time of the glycoproteins. In diseases such as cancer, marked variations in the composition of the carbohydrate portions of membrane glycoproteins and glycolipids have been observed. These alterations may result in the destruction of cell to cell recognition signals leading to loss of contact inhibition and increased cellular mobility.

VI. SOME ANTIBIOTICS ACTING AS INHIBITORS OF PROTEIN SYNTHESIS

Inhibitors of Transcription	Mode of Action
Actinomycin D	Binds to DNA in GC rich regions leading to inhibition of transcription, especially ribosomal RNA.
α-Amanitin	Inhibits eukaryotic RNA polymerases I, II, and III to varying degrees.
Cordycepin	Inhibits chain elongation during RNA transcription and poly(A) tail formation.
Rifampicin	Inhibits bacterial RNA synthesis by binding to RNA polymerase.

Inhibitors of Translation (Act at ribosomal level)

Bacterial

Tetracyclines	Prevent initial binding at entry site. Maximum inhibition of aa-tRNA binding is 50%, thus it appears to be specific for one site only.
Chloramphenicol	Acts at a step after binding and during peptide bond formation. It binds to the 50S subunit and stops the formation of a peptide bond. It also inhibits the puromycin-induced premature release of peptides.
Streptomycin and **dihydrostreptomycin**	Bind to 30S subunit of the ribosome causing distortion of the structural relationships between the ribosomes, mRNA, and aa-tRNA leading to misreading of the code. The products are non-functional proteins.
Erythromycin	Inhibits the translocation step during protein synthesis in bacteria.
Tetracycline	Blocks the ribosomal A site and inhibits protein synthesis in bacteria.

Mammalian

Cycloheximide, emetine and **dihydro-emetine**)	Inhibit protein synthesis in L-cells, reticulocytes, liver slices, yeast but do not affect bacteria. Exact site of action unknown, but seems to be near site of peptide bond formation. The rate of translation is slowed down markedly and the release of completed peptide chains is inhibited. Do not act on prokaryotic ribosomes.

Miscellaneous antibiotics

Diphtheria toxin	Covalently modifies elongation factor EF-2 in mammalian cells, leading to inhibition of protein synthesis.
Nalidixic acid and Novobiocin	Inhibit topoisomerases leading to inhibition of DNA synthesis in bacteria.
Puromycin	Mimics the aminoacyl region of aminoacyl tRNA and causes premature release of polypeptide chains during protein synthesis in bacterial and eukaryotic cells.
Tunicamycin	Inhibits N-linked glycosylation events in animal cells.

VII. REVIEW QUESTIONS ON PROTEIN BIOSYNTHESIS

DIRECTIONS: For each of the following multiple-choice questions (1 - 45), choose the ONE BEST answer.

---

1. If GGC is a codon in mRNA (5′-3′ direction), which one of the following would be the anticodon (5′-3′ direction) in tRNA?

A. GCC
B. CCG
C. CCC
D. CGC
E. GGC

2. Which one of the following compounds is used to initiate protein synthesis on mammalian ribosomes?

A. emetine
B. puromycin
C. formylated Met-tRNA
D. unformylated Met-tRNA
E. unformylated Arg-tRNA

3. Proteins destined for secretion in mammalian cells contain:

A. A hydrophilic "signal sequence" at their N-terminus.
B. Poly A at their C-terminus
C. A hydrophobic "signal sequence" at their N-terminus.
D. A "cap" structure at their N-terminus
E. Poly A at their N-terminus

4. Which one of the following antibiotics will inhibit protein synthesis in both prokaryotic and eukaryotic organisms?

A. cycloheximide
B. tetracycline
C. streptomycin
D. emetine
E. puromycin

5. tRNA molecules have at their 3′ termini the sequence:

A. CCA
B. CAA
C. CCC
D. AAC
E. AAA

6. The sequence of a short duplex DNA is

5′ GAACCTAC 3′
3′ CTTGGATG 5′.

What is the corresponding mRNA sequence transcribed from this region?

A. GAACCTAC
B. CUUGGAUG
C. CAUCCAAG
D. GAACCUAC
E. GUAGGUUC

7. The large subunit of mammalian ribosomes has a sedimentation constant of:

A. 40 S
B. 70 S
C. 30 S
D. 80 S
E. 60 S

8. A cluster of ribosomes translating the same mRNA is called a(n):

A. episome
B. monosome
C. polysome
D. spliceosome
E. genome

9. Most mammalian genes are "split", i.e., they contain coding and non-coding segments. After transcription into pre-mRNA the non-coding parts are removed or "spliced out". These segments are called:

A. spliceosomes
B. introns
C. exons
D. promoters
E. signal pieces

10. When the ribosome reaches one or more of the chain termination codons on the mRNA, release factors act to change the specificity of peptidyl transferase so that the growing peptide chain is transferred to:

A. water
B. EF-G
C. a membrane-bound polysome
D. the next incoming amino acid
E. the A-site

11. According to Blobel's hypothesis, all proteins destined for secretion contain a sequence of amino acids:

A. at their C-terminal ends that are primarily hydrophilic in behavior
B. at their C-terminal ends that are primarily hydrophobic in behavior
C. at their N-terminal ends that are primarily hydrophilic in behavior
D. at their N-terminal ends that are primarily hydrophobic in behavior
E. in the middle of the polypeptide chain that are primarily hydrophilic in behavior

12. Amino acid activating enzymes couple amino acids to the ribose moiety of the terminal adenylate residue of tRNA by forming an ester bond between the amino acid's alpha carboxyl and the ribose's:

A. $1'$ OH
B. $2'$ OH
C. $3'$ OH
D. $4'$ OH
E. $5'$ OH

13. In the asparagine-linked glycoproteins, the "mannose core" oligosaccharide is transferred from the cytosol, through the membrane of the endoplasmic reticulum and to the intra-luminal polypeptide chain by way of:

A. tocopherol phosphate
B. vitamin K phosphate
C. dolichol phosphate
D. isoprenol phosphate
E. vitamin A phosphate

14. Chain termination codons in mRNA are recognized by which proteins?

A. release factors
B. restriction enzymes
C. elongation factors
D. cap binding protein
E. initiation factors

15. Proteins are synthesized:

A. from the N-terminal to the C-terminal end
B. from the C-terminal to the N-terminal end
C. from rRNA
D. from 19 amino acids
E. in the nucleus

16. A point mutation in a mRNA codon will most likely cause:

A. mRNA degradation
B. inactivation of ribosomes
C. an altered amino acid sequence
D. incomplete transcription
E. inhibition of splicing

17. Which one of the following is NOT an example of post-transcriptional modification of mammalian hnRNA?

A. polyadenylation
B. splicing together of exons
C. degradation
D. capping
E. splicing together of introns

18. Diphtheria toxin inhibits:

A. protein synthesis initiation
B. protein synthesis elongation
C. protein synthesis termination
D. DNA supercoiling
E. mRNA splicing

19. The signal recognition particle recognizes:

A. RNA polymerase
B. DNA polymerase
C. nucleosomes
D. the N-terminus of secretory proteins
E. poly A

20. Each ribosome in a polysome is:

A. moving in the 3′ to 5′ direction on the mRNA
B. synthesizing many polypeptide chains
C. synthesizing only one polypeptide chain
D. dissociated
E. inhibited by actinomycin D

21. Small nuclear RNAs (snRNA) are involved in:

A. DNA synthesis
B. RNA synthesis
C. splicing
D. aminoacylation
E. recombination

22. The codon UUC is specific for the amino acid:

A. leucine
B. lysine
C. proline
D. phenylalanine
E. hydroxyproline

23 The genetic code is:

A. non-degenerate
B. species specific
C. a triplet code
D. punctuated
E. translated by RNA polymerase

24. A codon mutation resulting from a deletion of a nucleotide base results in a:

A. translocation
B. inversion
C. frameshift change
D. transversion
E. transition

25. Which one of the following statements about mammalian mRNA is NOT correct?

A. contains an unusual structure at its 5′ end
B. has a long stretch of poly A at its 3′ end
C. is transcribed as a hnRNA precursor
D. is transcribed by RNA polymerase I
E. has a relatively short half-life in prokaryotes

26. tRNA molecules have structural elements which recognize:

A. mRNA codons
B. aminoacyl synthetases
C. elongation factor Tu
D. ribosomal subunits
E. all of the above

27. The 70S ribosome contains about 50 separate proteins and one clearly definable enzymatic activity involved in peptide bond formation. The name of that activity is:

A. 30S dependent ATPase
B. aminoacyl-tRNA synthetase
C. elongation factor G
D. peptidyl transferase
E. IF-3

28. The majority of the peripheral sugars of the oligosaccharide side chains of complex glycoproteins are attached in the:

A. rough endoplasmic reticulum
B. smooth endoplasmic reticulum
C. lysosomes
D. Golgi membranes
E. plasma membranes

29 RNA synthesized in the nucleolus will most likely end up:

A. in eukaryotic chromatin
B. in prokaryotic ribosomes
C. as mRNA
D. in eukaryotic ribosomes
E. degraded to tRNA

30. The genetic code is degenerate because:

A. multiple species of ribosomes exist
B. multiple species of tRNA exist for most amino acids
C. there is a great inaccuracy during the process of transcription
D. a common codon exists for at least two amino acids
E. the code is not universal

31. Nascent glycoproteins whose mannose cores become phosphorylated are destined to be:

A. membrane proteins
B. secreted from the cell
C. transported to the nucleus
D. transported to lysosomes
E. degraded to amino acids

32. Which of the following compounds will be used to initiate eukaryotic protein synthesis?

A. Acetyl seryl tRNA.
B. Acetyl serine.
C. Formyl methionine
D. Formyl methionyl tRNA$_f$
E. Methionyl-tRNA$_i$

33. Which of the following factors are required for addition an amino acid to a growing polypeptide chain during protein biosynthesis in *E. coli?*

A. IF-1, IF-2 and IF-3
B. EF-G
C. R1
D. EF-Tu and EF-Ts
E. ATP

34. In eukaryotic cells, those proteins destined to be secreted from the cell are formed:

A. in the nucleolus
B. in the mitochondria
C. on free ribosomes
D. on membrane-bound polysomes
E. on free polysomes

35. Which of the following antibiotics binds to 70S ribosomal subunits, and if the concentration is high enough, will cause mis-reading of the genetic code?

A. cycloheximide
B. puromycin
C. chloramphenicol
D. dihydrostreptomycin
E. tetracycline

36. A point mutation resulting in a single base change in a mRNA codon will most likely cause:

A. inactivation of ribosomes
B. no effect
C. inactivation of IF-3
D. mRNA hydrolysis
E. an amino acid change in a polypeptide chain

37. An anticodon is:

A. the part of a DNA molecule which codes for chain termination
B. a 3-nucleotide sequence of a mRNA molecule
C. a specific part of a tRNA molecule
D. a nucleotide triplet of a rRNA molecule
E. the portion of a ribosomal subunit which interacts with the amino acid activating enzyme

38. During elongation, for each amino acid added to the growing polypeptide chain, the energy required per each ribosome is furnished by:

A. 1 GTP
B. 2 GTP
C. 3 GTP
D. 4 GTP
E. 1 ATP

39. Which of the following has a "poly-A tail"?

A. bacterial mRNA
B. eukaryotic mRNA
C. rRNA
D. DNA
E. bacterial tRNA

40. Eukaryotic ribosomes sediment with a value of:

A. 30S
B. 40S
C. 70S
D. 80S
E. 50S

41. All tRNAs have the same:

A. base sequence
B. base composition
C. axial length
D. anticodon
E. amino acid activating enzymes

42. Which of the following is a termination codon?

A. UAA
B. AUG
C. UUU
D. AAA
E. GCA

43. The 5'-"cap" on eukaryotic mRNA is essential for:

A. termination of translation
B. initiation of translation
C. transport of newly-formed mRNA from the nucleus
D. initiation of transcription
E. hydrolysis of GTP

44. An antibiotic that binds to the 50S subunit of a bacterial ribosome and blocks the A-site preventing the entrance of a new aminoacyl-tRNA complex is:

A. tetracycline
B. chloramphenicol
C. penicillin
D. dihydrostreptomycin
E. cycloheximide

45. When the signal recognition particle (SRP) is bound to the signal peptide on the growing polypeptide chain:

A. the elongation rate is accelerated
B. the elongation rate is decreased slightly
C. elongation is stopped completely
D. chain termination occurs
E. the signal peptide is cleaved off

MATCHING: For each set of questions, choose the ONE BEST answer from the list of lettered options above it. An answer may be used one or more times, or not at all.

Questions 46 - 49:

A. transfer RNA (tRNA)
B. ribosomal RNA (rRNA)
C. Both
D. Neither

46. In bacteria, is found as 16S and 23S complexes.

47. Forms ester bonds with activated amino acids.

48. Contains amino acid-specific codons.

49. Synthesis is inhibited by actinomycin D.

Questions 50 - 53:

A. EF-G
B. peptidyl transferase
C. Both
D. Neither

50. Forms peptide bond.

51. Inhibited by chloramphenicol in bacteria.

52. Requires GTP.

53. Participates in translocation of peptidyl tRNA.

VIII. ANSWERS TO QUESTIONS ON PROTEIN BIOSYNTHESIS

1. A	15. A	29. D	43. B
2. D	16. C	30. B	44. A
3. C	17. E	31. D	45. C
4. E	18. B	32. E	46. B
5. A	19. D	33. D	47. A
6. D	20. C	34. D	48. D
7. E	21. C	35. D	49. B
8. C	22. D	36. E	50. B
9. B	23. C	37. C	51. B
10. A	24. C	38. B	52. A
11. D	25. D	39. B	53. A
12. C	26. E	40. D	
13. C	27. D	41. C	
14. A	28. D	42. A	

14. RECOMBINANT DNA TECHNOLOGY

Jay Hanas

Introduction: The ability to detect and isolate normal and mutated genes, rapidly sequence them, and express large amounts of their protein products has had a tremendous impact on biology and medicine. This expertise, referred to as recombinant DNA technology, originated in the early 1970's with the discovery of *restriction endonucleases*, enzymes capable of specifically cleaving DNA into definable fragments. This chapter will describe this technology as well as more recent advances in this exciting field.

I. CONSTRUCTION AND CLONING OF RECOMBINANT DNA MOLECULES

A. Review of DNA/Gene Structure

Deoxyribonucleic acid (DNA) is composed of two anti-parallel ($5'$-phosphate → $3'$-hydroxyl, $3'$-hydroxyl → $5'$-phosphate) polynucleotide strands held together by hydrogen bonds between complementary dAMP-dTMP and dCMP-dGMP base pairs (bp). Genes are linear arrays of DNA base-pairs and are usually transcribed into single-stranded RNA (ribonucleic acid) molecules. A chromosome is a linear array (circular in bacteria) of genes and a genome is the haploid complement of DNA in the cell. A bacterial cell (e.g. *E. coli*) contains one chromosome (haploid) with about 3×10^6 bp and about 2000 genes. A human cell contains 46 chromosomes (diploid, 23 haploid pairs) with about 3×10^9 base pairs (haploid genome size) and about 50,000 genes.

Because of statistical randomness in any DNA sequence, every 4^4 (256) bp or 4^6 (4096) bp will on the average include, for example, the following unique sequences (or any other 4 or 6 bp sequence):

$$\begin{array}{ccc} \text{GATC} & & \text{GAATTC} \\ & \text{or} & \\ \text{CTAG} & & \text{CTTAAG} \end{array}$$

By convention, the top strand in a printed double-stranded DNA sequence is read $5'$-phosphate to $3'$-hydroxyl and the bottom strand is read $3'$-hydroxyl to $5'$-phosphate. The two double-stranded sequences illustrated above are called **palindromes** because they read exactly the same on both strands in the $5'$ to $3'$ direction.

B. Restriction Enzymes

Restriction enzymes are endonucleases which cleave (hydrolyze) in a sequence-specific manner $3'$-$5'$ phosphodiester bonds between adjacent base-pairs in double-stranded DNA. The most common type of restriction enzymes (type II) used in recombinant DNA technology come from bacteria and do not require an energy source for bond cleavage. They do require Mg^{++} ions for specific DNA binding and cleavage. Most restriction enzymes cleave either 4 or 6 bp DNA palindromes. Most restriction enzymes cleave a unique sequence. Those enzymes which recognize the same sequence are called **isoschizomers**. The type II restriction enzymes are part of a restriction/modification system in bacteria. For every specific restriction enzyme there is a methylating enzyme with the same DNA sequence specificity. For many of the restriction enzymes, prior methylation of their specific DNA site inhibits cleavage. This restriction/modification system is a defense mechanism allowing degradation of invading foreign DNA but protection of modified, cellular DNA.

In the 6 bp palindrome illustrated above, the restriction enzyme EcoRI (an enzyme isolated from *E. coli*), cleaves the phosphodiester bond between the G and A on both strands to yield two asymmetric fragments with $5'$-phosphate ends that overhang the $3'$-hydroxyl ends:

$$\begin{array}{ccc} \text{G-}3'\text{-OH} & & 5'\text{-P-AATTC} \\ & + & \\ \text{CTTAA-}5'\text{-P} & & 3'\text{-OH-G} \end{array}$$

Some restriction enzymes cleave in a symmetrical fashion (resulting DNA fragments have blunt ends) while others cleave asymmetrically resulting the 3' overhangs. On the average, a genome the size of the bacterium *E. coli* would have about 800 EcoRI cleavage sites and the human genome would have about 800,000 EcoRI cleavage sites. DNA fragments generated by restriction enzyme digestion can be resolved by agarose gel electrophoresis. This method takes advantage of the porosity of the agar-like agarose through which small DNA fragments migrate faster than larger fragments in the presence of a buffer solution and an electric field.

C. DNA Ligase

DNA ligase is an enzyme that catalyzes the formation of a phosphodiester bond between adjacent 3'-OH and 5'-phosphate termini in DNA. For activity, the enzyme requires Mg^{++} and a covalently bound AMP moiety, derived from ATP in the case of the DNA ligase of bacteriophage T4. The T4 enzyme can covalently join DNA molecules with compatible cohesive termini (for example hydrogen bonding of the 5' overhangs of EcoRI digested DNA fragments) and also blunt-ended DNA fragments. *The use of DNA restriction digestion followed by covalent joining of cohesive ends from different DNA fragments forms the basis of recombinant DNA enzymology.*

D. Plasmids and Cloning of Foreign DNA

Plasmids are small DNA molecules (about 5000 bp) that replicate autonomously in bacterial cells. They are covalently closed circles of double-stranded DNA. Most plasmids contain antibiotic resistance genes which allow growth of bacterial cells harboring the plasmids under selective conditions. Examples of antibiotic resistance genes are those coding for beta-lactamase or neomycin phosphotransferase, enzymes that inactivate ampicillin or aminoglycosides, respectively.

Because of their relatively small size, plasmids contain very few unique restriction enzyme cleavage sites. Upon digestion of a plasmid at a unique restriction site, the circular plasmid is rendered linear. This DNA molecule, referred to as a vector, can now be ligated with a foreign DNA fragment (insert) containing the same cohesive termini as the vector (digested with the same restriction enzyme). Ligation by DNA ligase forms a recombinant DNA molecule with a covalently closed structure. A circular recombinant molecule can now enter a bacterial cell, replicate, and, by virtue of the plasmid antibiotic resistance gene, transform the normally sensitive cell to one that will grow in the presence of that antibiotic.

If the ligation is performed in a test tube with millions of EcoRI- digested plasmid molecules and millions of unique foreign DNA fragments with cohesive termini (e.g EcoRI digested human genome), then millions of recombinant DNA molecules are formed, each containing the same plasmid vector DNA but with a unique fragment (gene, in some cases) of human DNA. When recipient bacterial cells are transformed with this mixture and platted on agar plates containing the proper antibiotic, only a cell containing a plasmid will grow into a visible colony containing millions of bacteria all containing the identical plasmid vector and foreign DNA insert (clones). Over a small number of agar plates, it is possible to have enough individual bacterial clones so that every gene in the human genome is cloned in a single bacterial colony. Such a colony collection is called a **genomic library** or **clone bank**.

Another library commonly used in recombinant DNA technology is a **cDNA** (complementary DNA) library. The foreign DNA inserts cloned in this library are double-stranded DNA molecules synthesized by the enzyme *reverse transcriptase* from the messenger RNA template population found in a specific tissue. Thus, a cDNA library contains cloned coding regions for all the proteins expressed in a particular tissue and will vary from tissue to tissue. The cloned DNA will not contain introns (unlike cloned genomic DNA) and therefore allows direct expression of the transcribed mRNA into protein *in vitro* or in bacteria.

II. SCREENING AND EXPRESSION OF FOREIGN DNA IN BACTERIAL CLONES.

A. Screening Clone Banks

Trying to pick out a specific colony harboring a specific gene or cDNA inserted in a plasmid out of several million present in a clone bank is akin to the proverbial search for the needle in a haystack. It is necessary to know something about the DNA fragment to be isolated. In most cases, this involves having the DNA sequence of an evolutionarily related gene, knowing the protein sequence encoded by the gene to be isolated, or having a specific antibody against this protein.

Screening is often performed by selective hybridization with a DNA probe. Hybridization refers to the process by which base-pairing or annealing occurs between complementary, single-stranded DNA strands. The probe can be a DNA fragment of several hundred or more base pairs with homology to the gene to be isolated. Also a small single-stranded DNA probe can be made synthetically which will hybridize with the genetic code for a known amino acid sequence. After transfer of a bacterial colony library from one or more petri dishes onto a paper or nylon filter, the bacterial debris is removed leaving the cellular and plasmid DNA bound to the filter in precisely the same position as the original bacterial colony (a replica of the original bacterial plate is retained). The DNA on the filter is then denatured and hybridized with denatured radioactively labeled DNA probe. Colonies containing DNA with sufficient homology to the radioactive probe "light up" when the washed, hybridized filter is exposed to X-ray film. The plasmid DNA containing the gene of interest is then purified from the original bacterial colony.

Screening gene libraries can also be accomplished using a specific antibody that will recognize and bind to the protein whose gene is to be isolated. The antibody can either be against the entire protein or can be directed against a synthetic peptide derived from the primary sequence of the protein. The antibody can be radioactively or enzymatically labeled and used to screen bacterial clone banks in which the foreign genes have been cloned into a special expression plasmid (described below). In this plasmid the inserted gene is transcribed into mRNA and translated into protein by the bacterial RNA polymerase and ribosomes. The bacterial colonies on the filter paper are lysed so the expressed protein in a specific colony can be detected by the antibody.

B. Expression of Foreign Genes in Heterologous Systems

The ability to express foreign genes in bacteria, yeast, or mammalian cells is important in screening for desired gene products and also for producing large amounts of medically important recombinant proteins. These include interferons, human growth hormone, colony stimulating factors, erythropoietin, clot busters (streptokinase, tissue plasminogen activator) and many more. Because bacteria can not process RNA transcribed from split genes (containing introns), they can only express cDNA which codes directly for a protein sequence. The cDNA is inserted into a bacterial plasmid between signals for proper initiation (promoter) and termination of transcription by bacterial RNA polymerase. Also, a sequence for initiation of translation (Shine-Delgarno sequence) by bacterial ribosomes is necessary downstream from the promoter. It may be important to regulate transcription of the foreign cDNA in *E. coli* for example by using the promoter/operator region from the lac operon. The expressed protein can also be engineered to be secreted out of the bacterial cell if a signal sequence is attached to the N- terminus of the protein.

One human protein expressed in and isolated from bacteria is insulin. Normally in pancreatic cells insulin is synthesized as a precursor (preproinsulin containing a leader sequence and other polypeptide regions which are cleaved off leaving the A-B chains held together by disulfide bonds. Synthesis of the recombinant protein in *E. coli* involves expression and purification of the A and B chains independently followed by re-assembly of the active A-B protein.

Gene screening and expression can also be accomplished in mammalian cells in culture. Eukaryotic cDNA or genomic DNA can be cloned in expression plasmids that function in eukaryotic cells. Such plasmids must contain eukaryotic transcription, RNA processing, and translation regulatory regions. The transfer of exogenous DNA into eukaryotic cells is termed **DNA transfection**.

III. OTHER IMPORTANT RECOMBINANT DNA TECHNIQUES

A. DNA Sequencing

The ability to sequence large pieces of DNA rapidly has revolutionized biology. The United States government will fund research over the next twenty years to map and sequence all 3×10^9 bp of the human genome (human genome initiative). Currently two basic methodologies exist to sequence DNA, both developed in the mid to late 1970's. These methods are (1) chemical degradation of DNA fragments, and (2) chain termination reactions during template-directed DNA synthesis. Because the chain termination method is widely used and much more amenable to sequencing large stretches of DNA rapidly, only its principles will be discussed.

DNA fragments to be sequenced are cloned into a bacteriophage genome which exists as a circular double-stranded DNA molecule during replication but is single-stranded otherwise. Single-stranded phage DNA containing one of the strands of the foreign DNA to be sequenced is subjected to *in vitro* DNA synthesis using DNA polymerase and a specific synthetic oligonucleotide as a hybridization primer which binds to the bacteriophage template just upstream of the inserted foreign DNA. In vitro DNA synthesis is conducted in four separate reaction vessels each containing all four dNTPs (one radioactive) plus a small amount of only one dideoxynucleotide (ddGTP, ddATP, ddCTP, or ddTTP). Unlike a deoxynucleotide which contains a sugar $2'$-H and $3'$-OH, the dideoxynucleotides contain hydrogens at both the $2'$- and $3'$-positions. Therefore, once they are incorporated into the growing DNA chain during synthesis ($5'$-phosphate added to $3'$-OH), they terminate synthesis of that chain because they lack the $3'$-OH required for addition of the next $5'$-phosphate dNTP. Because the dideoxynucleotides are present in small amounts relative to the dNTPs in the individual reactions, chain termination is random and results in a very wide range of newly synthesized chain sizes (20 to 1000 b). Because acrylamide gel electrophoresis is capable of resolving the individual chains that differ by 1 base, all four reaction mixtures are subjected to electrophoresis through polymerized acrylamide in adjacent lanes. Upon completion of electrophoresis, the gel is exposed to X-ray film where the radioactive chain termination products are revealed as bands of differing electrophoretic mobility. The newly synthesized sequence is read in one-base increments across the four reaction lanes (ddG, ddA, ddT, ddC) from bottom to top, the $5'$ to $3'$ direction. The complement of this sequence will be the sequence of the foreign DNA template in the $3'$ to $5'$ direction.

B. Restriction Fragment Length Polymorphism

Except in identical twins, there exists a significant amount of DNA sequence variation between individuals. Identical genes in different individuals (not limited to humans) may differ in length between the same restriction enzyme cleavage site. This phenomenon is called **restriction fragment length polymorphism (RFLP)**. RFLPs can be due to nucleotide base changes (point mutations, additions, deletions) causing the loss of an existing restriction site or gain of a new one or due to the presence of variable numbers of tandem repeats (VNTRs, small repetitive DNA elements) between two restriction sites. The functional significance of RFLPs is not known although gross differences in DNA structure can have pathological effects (e.g. gene translocations). Several hundred human RFLPs have already been identified and many more exist. Because autosomal RFLPs are inherited in Mendelian fashion (one allele from each parent), genes responsible for many genetic diseases, e.g. cystic fibrosis, can be traced and identified in affected families by RFLP analysis.

This process is also referred to as reverse genetics because a specific gene is isolated in the absence of a known protein or function.

Experimentally, RFLP analysis is performed by isolating genomic DNA from tissue samples, digesting the DNA with different restriction enzymes, and then electrophoresing the DNA through an agarose gel. The DNA bands are then transferred by capillary action onto nitrocellulose paper, preserving the gel migration pattern (smaller fragments further down the gel than larger fragments). This transfer of DNA from the gel to paper is referred to as **Southern blotting** or **Southern transfer**. Transfer of RNA or protein from a gel to paper is called a **Northern** or **Western** blot respectively. The DNA on the Southern blot is denatured and then hybridized with different probes, containing either unique or repetitive DNA sequences. The DNA probes are either labeled isotopically or with a non-radioactive substrate which is subsequently identified enzymatically. Variable length fragments in different individuals generated by the same restriction enzyme will be observed if an RFLP is present. The identification of a gene responsible for a genetic disease begins with detection of a unique RFLP fragment or fragments that have the same familial pattern of inheritance as the disease itself. This method has been used to positively identify the aberrant genes responsible for cystic fibrosis and muscular dystrophy among others.

RFLPs are also very useful in forensic medicine. Biological evidence left at a crime scene (blood, skin, semen, hair) is a source of DNA which can be subjected to RFLP analysis. Because no two individuals will have identical RFLP patterns (DNA fingerprints) at multiple genetic loci (except identical twins), victim, perpetrator, or suspect DNA (victim or suspect DNA is easily obtained from a blood sample) can be identified and compared. Suspects can therefore be implicated or absolved.

C. Polymerase Chain Reaction

The **polymerase chain reaction** (PCR) was developed in the mid-1980's in order to amplify minute amounts of nucleic acid. It has special usefulness in gene cloning and medical diagnostics. In PCR, a double-stranded DNA fragment is denatured by heating into its two complementary strands. Complementary single-stranded DNA oligonucleotides are then hybridized to the respective ends of the denatured strands of the DNA fragment upon cooling. The $3'$-OH ends of these hybridized oligonucleotides serve as primers for DNA synthesis by a thermostable DNA polymerase, resulting in synthesis of two daughter fragments. Repeated denaturation, hybridization with oligonucleotide, and DNA synthesis cycles allow huge amplification to occur. For example, 20 such cycles will result in a million-fold amplification and 30 cycles will result in a billion-fold amplification. Genetic sequence of RNA can also be amplified if a cDNA strand is initially polymerized by reverse transcriptase. PCR can amplify and therefore identify a single AIDS virus in a ml of blood or amplify DNA from a single sperm cell.

D. Other Relevant Molecular Biology Techniques

In the Southern blotting technique, DNA is subjected to electrophoresis and then transferred to paper or nylon. RNA and protein can be analyzed in similar fashion. Electrophoresis of RNA through agarose or acrylamide followed by transfer to paper or nylon is called Northern transfer. Transferred RNA can be detected by hybridization with a radioactive probe. Northern blotting reveals the pattern of gene expression at the RNA level in cells or tissues. Electrophoresis of denatured proteins through acrylamide followed by transfer to paper is called Western blotting. The transferred protein can then be detected most easily by immunological methods. For example, a specific rabbit antibody can bind its protein antigen on the paper. Another antibody (anti-rabbit) conjugated to an enzyme (alkaline phosphatase) can then bind the original rabbit antibody on the paper. When the paper is then soaked in a substrate stain, a dark precipitate will deposit over the antigen band. Western blotting is used to detect the presence of HIV antigens or HIV antibodies in human serum as well as other viral antigens.

Another electrophoresis technique that is useful in the mapping and sequencing of large genomes is **pulse-field gel electrophoresis.** In this method, very large DNA molecules (millions of base-pairs, chromosome size) are separated by size on agarose gels. The electrophoretic field in which the DNA molecules migrate is not continuous or uniform. The field is pulsed on and off and is bidirectional. This results in the slow but progressive migration of very large molecules through the agarose pores without shearing. Thus, individual chromosomes can be resolved and isolated by this technique.

In situ **hybridization** is a process by which nucleic acid probes are hybridized to complementary nucleic acids inside of fixed cells and tissue slices (*in situ*). If the probe is radiolabeled with tritium, the hybridized sample can be exposed to X-ray film to determine whether the cells/tissue contain complementary nucleic acids. This method can be used for diagnostic purposes by using probes specific for viral nucleic acids, altered genes, or oncogenes.

Anti-sense nucleic acids can be used to inhibit virus and cell gene expression. By convention, the top strand of a printing double-stranded DNA sequence is termed the coding or sense strand. This top strand will have the same sequence or sense as the RNA transcribed off the bottom template strand (non-coding or anti-sense). Therefore, the presence of an anti-sense nucleic acid that will hybridize to the translation start region of a specific mRNA will inhibit ribosome binding and translation of this mRNA. The anti-sense nucleic acid can be introduced to the cell directly or can be expressed off a transfected cellular plasmid.

IV. REVIEW OF TERMINOLOGY

The student should know the meaning and significance of the following:

restriction enzyme	RFLP	DNA transfection
plasmid	PCR	DNA fingerprinting
DNA ligase	Southern blot	Reverse genetics
gel electrophoresis	Northern blot	pulse-field electrophoresis
genome	Western blot	recombinant protein
gene	anti-sense nucleic acids	human genome project
chromosome	in situ hybridization	chimeric plasmid
palindrome	probe	reverse transcriptase
clones	hybridization	cDNA
dideoxynucleotides	DNA sequencing	clone bank
antibiotic resistance	VNTR	chain termination
genes	haploid	autoradiography

V. REVIEW QUESTIONS ON RECOMBINANT DNA TECHNOLOGY

> **DIRECTIONS:** For each of the following multiple-choice questions (1 - 24), choose the ONE BEST answer.

1. Southern blotting techniques involve:

A. the transfer of DNA fragments from a gel to a membrane support
B. the hybridization of a labeled probe to the immobilized DNA
C. the detection of the immobilized fragment which has nucleic acid sequence homology to the probe
D. all of the above
E. none of the above

2. You wish to locate a Protein Z clone in a cDNA library cloned into an expression vector. You do not know the nucleotide sequence of the gene or the amino acid sequence of the protein. What tool do you need to isolate the clone?

A. an antibody recognizing Protein Z
B. a liver cDNA library
C. the hormone regulating the expression of protein Z
D. a synthetic oligonucleotide

3. Transposable elements have been shown to be agents causing mutations in humans. This observation was based on:

A. the mobility of transposable elements when exposed to UV light
B. the deletion of DNA in the mutant gene
C. the formation of direct repeats flanking the insertion of new DNA
D. the detection of transposase activity in the mutant individual

4. You have obtained a DNA fragment from another researcher which represents a portion of the vinculin gene. You carry out a Northern blot from liver tissue, which tells you:

A. the intron processing of the vinculin transcript
B. the size and abundance of the vinculin transcript
C. the size of the vinculin gene
D. the number of vinculin genes in the genome

5. Retroviruses are being used as vectors for the first gene therapy experiments. What properties of retroviruses have made them appropriate for this use?

A. the ability to insert into the genome
B. the ability to cause transcription of the genes they carry
C. the ability to infect cells
D. all of the above

6. You are cloning BamHI-cut genomic DNA into the plasmid vector pBR322, which has both an ampicillin resistance gene and a tetracycline resistance gene. If the inserts are ligated into the BamHI site in the tetracycline resistance gene of the vector, the entire population of transformed bacteria should first be plated onto:

A. agar plates containing no antibiotic
B. agar plates containing ampicillin
C. agar plates containing tetracycline
D. agar plates containing both ampicillin and tetracycline

7. In the protocol for DNA sequence analysis developed by Maxam and Gilbert, partial chemical cleavages generate ladders of labeled DNA fragments. If you are carrying out this kind of analysis using DNA fragments labeled at their 3′ ends, as you read the DNA sequencing gel from the smaller fragments to the larger fragments, you will be reading:

A. 5′ to 3′ from bottom to top of gel
B. 3′ to 5′ from bottom to top of gel
C. 5′ to 3′ from top to bottom of gel
D. 3′ to 5′ from top to bottom of gel

8. The restriction site for EcoRI is:

A. GAATTC
   CTTAAG

B. GGAACC
   CCTAGG

C. AAGGTT
   TTCGAA

D. AGATCC
   TCTAGG

E. AAGCTC
   TTCGAG

9. An example of a DNA palindrome is:

A. AGATCC
   TCTAGG

B. AGGTCT
   TCCAGA

C. GGATCT
   CCTAGA

D. AAGCTC
   TTCGAG

E. AGATCT
   TCTAGA

10. Restriction fragment length polymorphism (RFLP) is a method to:

A. amplify DNA
B. identify individuals
C. regulate gene expression
D. degrade lactose
E. sequence DNA.

11. DNA hybridization is:

A. melting of duplex DNA
B. annealing of complementary strands of DNA
C. covalent closing of a duplex DNA molecule to give a circle
D. DNA degradation
E. DNA sequencing.

12. Compared to the genome of the bacterium *E. coli*, the human genome is about:

A. the same size
B. 10 times larger
C. 1000 times larger
D. a million times larger
E. composed of RNA

13. DNA ligase:

A. cleaves DNA at specific sites
B. covalently joins DNA fragments
C. converts DNA into RNA
D. replicates DNA.
E. is involved in RNA splicing

14. Plasmids are:

A. linear DNA molecules
B. circular DNA molecules
C. not replicated in *E. coli*
D. photosynthetic organelles.
E. examples of cDNA

15. Most plasmids contain:

A. a ribosome binding site
B. a lac repressor
C. a lac operator
D. an antibiotic resistance gene
E. a single DNA strand

16. The best definition of a gene is:

A. a piece of double-stranded DNA
B. a piece of single-stranded DNA
C. DNA that is transcribed into RNA
D. DNA that contains nucleosomes.
E. DNA that codes for a protein

17. Polymerase chain reaction (PCR) is used to:

A. sequence DNA
B. sequence protein
C. stimulate RNA polymerase
D. degrade transcription factors
E. amplify a gene.

18. Restriction enzymes cleave specific sequences:

A. in single stranded DNA
B. in single stranded RNA
C. in double stranded RNA
D. in double stranded DNA
E. in denatured DNA

19. Bacteriophage T4 DNA ligase requires for activity:

A. GTP
B. CTP
C. a palindrome
D. UTP
E. ATP

20. DNA sequencing by the chain-termination method requires the use of:

A. dideoxynucleotides and DNA polymerase
B. dideoxynucleotides and restriction enzymes
C. ATP
D. DNA ligase
E. plasmids

21. The first step in PCR is to:

A. perform the DNA polymerization reaction
B. hybridize primers to the single stranded DNA templates
C. electrophoresis of DNA on an agarose gel
D. perform Southern blotting
E. expose X-ray film

22. Agarose and polyacrylamide gel electrophoresis separate DNA molecules on the basis of:

A. source of DNA
B. concentration of DNA
C. molecular mass (size)
D. base composition
E. charge

23. All students (except identical twins) will have different

A. chromosome numbers
B. RFLP patterns
C. genome sizes
D. ribosomal gene number
E. DNA polymerases.

VI. ANSWERS TO QUESTIONS ON RECOMBINANT DNA TECHNOLOGY

1.	D	9.	E	17.	E
2.	A	10.	B	18.	D
3.	C	11.	B	19.	E
4.	B	12.	C	20.	A
5.	D	13.	B	21.	B
6.	B	14.	B	22.	C
7.	B	15.	D	23.	B
8.	A	16.	C		